SpringerBriefs in Molecular Science

More information about this series at http://www.springer.com/series/8898

Yun-Pei Zhu · Zhong-Yong Yuan

Mesoporous Organic-Inorganic Non-Siliceous Hybrid Materials

Basic Principles and Promising Multifunctionality

 Springer

Yun-Pei Zhu
College of Chemistry
Nankai University
Tianjin
China

Zhong-Yong Yuan
College of Chemistry
Nankai University
Tianjin
China

ISSN 2191-5407   ISSN 2191-5415   (electronic)
SpringerBriefs in Molecular Science
ISBN 978-3-662-45633-0   ISBN 978-3-662-45634-7   (eBook)
DOI 10.1007/978-3-662-45634-7

Library of Congress Control Number: 2014955613

Springer Heidelberg New York Dordrecht London

© The Author(s) 2015
This work is subject to copyright. All rights are reserved by the Publisher, whether the whole or part of the material is concerned, specifically the rights of translation, reprinting, reuse of illustrations, recitation, broadcasting, reproduction on microfilms or in any other physical way, and transmission or information storage and retrieval, electronic adaptation, computer software, or by similar or dissimilar methodology now known or hereafter developed.
The use of general descriptive names, registered names, trademarks, service marks, etc. in this publication does not imply, even in the absence of a specific statement, that such names are exempt from the relevant protective laws and regulations and therefore free for general use.
The publisher, the authors and the editors are safe to assume that the advice and information in this book are believed to be true and accurate at the date of publication. Neither the publisher nor the authors or the editors give a warranty, express or implied, with respect to the material contained herein or for any errors or omissions that may have been made.

Printed on acid-free paper

Springer-Verlag GmbH Berlin Heidelberg is part of Springer Science+Business Media (www.springer.com)

Preface

Nowadays, the development of new materials is focusing on the rational design of advanced systems with particular properties that can be predicted and controlled, aiming at pre-determined technological applications. This suggests that the research within the field of Materials Science should be based on the criteria of multidisciplinarity, thereby allowing the design and preparation of specific materials. Organic–inorganic hybrid materials constitute indeed a significant and promising category within the realm of Materials Science. The infinite kinds of organic functional groups, judicious control of inorganic units, and their corresponding marvelous assemblies endow them with tremendous potential to yield new materials beyond conventional composites, a domain in which nanocomposites push forward the frontier of discovery and advanced functional materials. Furthermore, the introduction of mesoporosity and even hierarchical porosity into the hybrid frameworks extends their application from traditional fields to high-tech areas. As a consequence, the encounter of hybrid chemistry and porous structures offers great opportunities for the development of functional materials, a fertile ground to harness the physicochemical, electrochemical, or biological activity of a myriad of organic and inorganic components and put them to scientific research and finally practical applications.

Providing a thorough list of contents that could fairly represent the large and fascinating family of porous organic–inorganic hybrid materials would be impossible. Instead, we have striven to present some emerging types including metal phosphonates, carboxylates, and sulfonates, exampling as the non-siliceous organic–inorganic hybrid materials, which would criss-cross the field revealing in some detail the basic principles and a variety of functional properties and applications. This book consists of six chapters. The Introduction (Chap. 1) describes the classification of porous materials. For better understanding of hybrids, Chap. 2 exhibits the development history of hybrid materials and strategies for integrating organic and inorganic moieties. The synthesis pathways of mesoporous non-siliceous hybrid materials and the key factors such as precursors, surfactants, adjustment of mesostructures and pore size, crystallization improvement of pore wall, as well as morphology control are elaborated in Chaps. 3 and 4, attempting to provide insights into synthesizing high-quality mesoporous non-siliceous hybrid materials. In Chap. 5, the applications of mesoporous hybrid materials and discussions of structure–function relationship are presented. It is apparent that the mesoporous hybrids field is eager for more and more researchers from various fields to explore attractive applications. Finally, the latest progresses in development of mesoporous non-siliceous organic–inorganic hybrid materials are reviewed, and the outlook on next stages is given.

Looking toward the twenty-first century, nanoscience and nanotechnology will make a significant contribution to scientific and technological development. Hybrid materials are believed to play a major role in the design and preparation of advanced functional materials. Recently, the molecular approaches in chemical synthesis and nanochemistry have reached a high level of sophistication. The synthesis of mesoporous hybrid materials is considerably promising to be mastered.

We hope that this book can help and inspire those researchers who are interested in porous hybrid materials. Due to the relatively wide area covered in this book and the limited knowledge and competence of the authors, errors and omissions may not be avoided, therefore we sincerely appreciate the criticism and comments from the readers.

Acknowledgments

The authors thank the support from NSFC, and the Program for Innovative Research Team in University.

Yun-Pei Zhu
Zhong-Yong Yuan

Contents

Chapter 1
Introduction

Abstract Porous materials with organic–inorganic hybrid framework are of great interest to many scientific communities. Rational design of mesoporous hybrid materials with specific functionalities is of fundamental importance. In this part, after a brief description of the mesoporous materials, the conception of organic–inorganic hybrid materials and the limitation of periodic mesoporous organosilicas are presented, followed by the protrusion of mesoporous non-siliceous organic–inorganic hybrids including metal phosphonates, sulfonates and carboxylates.

Keywords Porous materials · Mesoporosity · Organic–inorganic hybrid · Metal phosphonates · Metal sulfonates · Metal carboxylates

As one of the most significant subjects of science and technology, materials science has boomed with the rapid development of human civilization. Where the core ideology lies is the invention and development of advanced multifunctional materials, which is essential for alternative and renewable sources, and the abatement of harmful substances. In particular, new materials not only greatly promote the advances of industry, agriculture, medicine, and information science, but also present some revolutionary transformation of the forms and novel functionalities, thereby resulting in enormous changes to human life.

The key point for designing new materials is to adjust the nature and accessibility of the inner interfaces. Porous materials, as a subset of nanostructured materials, possess the ability to interact with atoms, ions, molecules, and even larger guest molecules, not only at the surface but also throughout the bulk of the material [1]. According to the International Union of Pure and Applied Chemistry (IUPAC) convention [2], porous materials are divided into three types on the basis of the pore size: microporous materials with pore size smaller than 2 nm, mesoporous materials with pore size ranging from 2 to 50 nm, and macroporous materials, where the size of pores are larger than 50 nm. The distribution of shapes and sizes of the void spaces in nanoporous materials is intimately related to their capability to perform a desired function in a particular area. Classical microporous zeolites have a uniform sieve-like pore structure and high specific surface

© The Author(s) 2015
Y.-P. Zhu and Z.-Y. Yuan, *Mesoporous Organic-Inorganic Non-Siliceous Hybrid Materials*, SpringerBriefs in Molecular Science,
DOI 10.1007/978-3-662-45634-7_1

area, thus exhibiting broad applications in chemical, petrochemical, gas separation industries, and other fields. For example, they can be employed to separate molecules based on the pore size by selectively adsorbing smaller molecules from a mixed system containing molecules too large to enter into the pores. However, the pore sizes of zeolites are in the microporous region, usually less than 1.3 nm, prohibiting the further applications that involve transfer and conversion of macromolecules. Accordingly, the exploration and rational design of porous materials with considerable porosity has become an important branch of materials science.

In comparison with microporous and macroporous materials, mesoporous materials have attracted more and more research interest and have shown great potentials in many areas due to their outstanding properties, such as approximate pore diameters, high surface areas, tunable porosity, alternative pore shape, and abundant compositions [3–5]. Mesoporous materials became a hot research topic in 1992, when Mobil Oil Corporation (Mobil) scientists first reported the M41S series of mesoporous silica materials [6, 7]. Cationic surfactants with long-chain alkyl were utilized as a structure-directing agent to prepare ordered mesoporous (alumino-) silicate materials. Furthermore, Mobil researchers not only developed a family of mesoporous materials with ordered pore arrangements, but also proposed a general "liquid-crystal templating" mechanism with detailed synthesis rules, thereby a new research area of inorganic synthetic chemistry began to rise. Scientific workers have witnessed a rapid development in mesoporous materials with new mesostructures and compositions. If we refer to ISI Web of Knowledge and use "mesoporous" as the subject, a predominantly increasing number of publications can be obviously observed (Fig. 1.1), exhibiting the emerging development trends in this field.

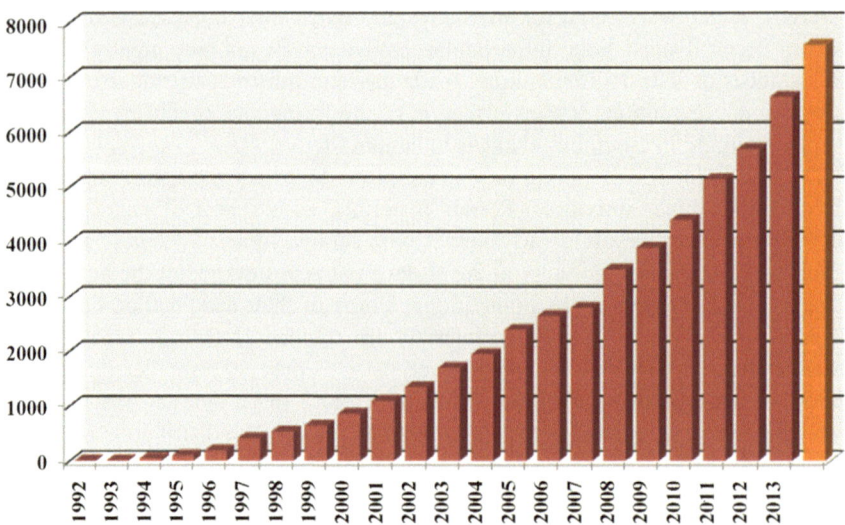

Fig. 1.1 Statistics of the published papers related to "mesoporous" according to ISI Web of Knowledge

To date, a great diversity of mesoporous materials has been synthesized. Since the first report of mesoporous silica materials (M41S) pioneered by Mobil Corporation, the most intriguing mesoporous silica-based materials, so-called SBA series, were explored by the Stucky group from the University of California, Santa Barbara. Besides, scientists from other countries, including China, Japan, Korea, Canada, UK, and France, have contributed a lot to the development of mesoporous families. The KIT series developed by a Korea research team, Prof. Ryoo's group in KAIST, and the FDU series developed by Prof. Zhao's group in Fudan University have been extensively investigated.

On the other hand, the use of pure inorganic silica-based materials is limited to physical properties concerned with catalytic supports and adsorbents, showing insufficiency with respect to the low mechanical strength and the difficulties in post-modification or functionalization. Interestingly, pure porous organic frameworks, such as covalent organic frameworks (COFs), represent an emerging class of porous polymers that have received tremendous interest for diverse applications, including chemical separations, gas storage, catalysis, and optoelectronic and charge storage devices [8–14], which is because the periodic organic building units offer the potential for judicious adjustment and post-functionalization. It should be kept in mind that organic networks show inferior thermal and chemical stability as compared to inorganic counterparts, and the poorly understood simultaneous polymerization and crystallization processes further inhibit the corresponding applications.

Deliberate efforts to combine the favorable properties of inorganic units and organic moieties in a single composite material represent an old challenge that started the beginning of the industrial era. The intimate integration of organic and inorganic components to form organic–inorganic hybrid materials constitutes indeed a remarkable and growing category within the realm of materials science. Numerous new applications involving advanced materials science are intimately related to functional hybrids. Accordingly, the combination at the nanosize level of active inorganic and organic or even bioactive components in a single material has made accessible an immense new area of materials science that has extraordinary implications in the development of multifunctional advanced materials. With the establishment of "*chimie douce*", Livage opened the gates toward a new galaxy of materials, organic–inorganic hybrid materials [15–18]. Later on, research turned toward more sophisticated nanocomposites with higher added values. Noticeably, the concept of "organic–inorganic hybrid materials" has more to do with chemistry than with simple physical mixtures. In general, organic–inorganic hybrid materials are nanocomposites with the inorganic constituents and organic components interacting intimately at the molecular scale [19, 20]. Nowadays the field of organic–inorganic materials has been extended to other fields as diverse as molecular and supramolecular materials or polymer chemistry. Furthermore, due to the combined physico-chemical merits of organic and inorganic components, a very significant trend is the growing research interest toward functional hybrids, which further broadens the field.

Periodic mesoporous organosilicas (PMOs) containing organic siloxane groups in the silica network have received much attention since 1999 [21–23]. The predominant process in the formation of siliceous hybrid mesoporous materials during the solgel procedure is the incorporation of organic groups via hydrolysis and polymerization using organically modified silanes. The homogeneous distribution of organic bridges both in the wall and on the surface is fascinating and valuable from the viewpoint of materials and chemistry. However, besides the limited choice and high cost of the precursors of organosilicon reagents, functionalization of organosilicas is confined to the physical properties concerned with adsorption, ion-exchange, and catalysis. The exploitation of hybrid materials has thus been extended to non-siliceous organic–inorganic hybrid materials.

Chemically designed non-siliceous organic–inorganic mesoporous hybrids, in which metal sulfonates, carboxylates, and phosphonates represent the three members of the family, are considered to be promising candidates for environmentally friendly and multifunctional materials [24, 25]. Different dimensions and reactivities of the bridging molecules lead to the distinct structures and stabilities of the resultant hybrids. Because of the variety of available organic acid linkages and their derivatives and the various metallic precursors, the physicochemical properties of hybrid frameworks can be designed and further modified adequately through the use of different metal ions and organic bridging molecules. Not only the hydrophobic/hydrophilic and acid/alkaline natures of the pore surface could be adjusted, but also the homogeneous introduction of functional binding sites into the framework could be realized. By ingenious selection of the synthesis systems and technology, the pore width, mesophase, crystallization of the pore walls, and morphology of the mesoporous non-siliceous hybrids can be effectively controlled. Thus, mesoporous metal–organic hybrids have been widely utilized in adsorption, separation, catalysis, photochemistry, and biochemistry, owing to their high surface area, large pore volume, adjustable porosity, easy handling, low-cost manufacturing, and intriguing surface properties [26–28].

This book summarizes and highlights the progress of mesoporous non-silica-based hybrid materials with controllable compositions and structural properties in the past decade, including metal sulfonates, carboxylates, phosphonates, and some "hot" MOFs. The purpose of this book is to provide a comprehensive review and, together with it, a future outlook. We hope that this book can be a good reference for a wide readership, including researchers, scientists, and students in chemistry, chemical engineering, physics, materials science, and biology, who are interested in mesoporous non-siliceous hybrid materials.

References

1. M.E. Davis, Ordered porous materials for emerging applications. Nature **417**, 813–821 (2002)
2. K.S.W. Sing, D.H. Everett, R.H.W. Haul, L. Moscou, R.A. Pierotti, J. Rouquerol, T. Siemieniewska, Physical and biophysical chemistry division commission on colloid and surface chemistry including catalysis. Pure Appl. Chem. **57**, 603–619 (1985)
3. M.B. Park, Y. Lee, A.M. Zhang, F.S. Xiao, C.P. Nicholas, G.J. Lewis, S.B. Hong, Formation pathway for LTA zeolite crystals synthesized via a charge density mismatch approach. J. Am. Chem. Soc. **135**, 2248–2255 (2013)

4. S. Che, Z. Liu, T. Ohsuna, K. Sakamoto, O. Terasaki, T. Tatsumi, Synthesis and characterization of chiral mesoporous silica. Nature **429**, 281–284 (2004)
5. X. Bu, P. Feng, G.D. Stucky, Isolation of germanate sheets with three-membered rings: a possible precursor to three-dimensional zeolite-type germinates. Chem. Mater. **11**, 3423–3424 (1999)
6. C.T. Kresge, M.E. Leonowicz, W.J. Roth, J.C. Vartulli, J.S. Beck, Ordered mesoporous molecular sieves synthesised by a liquid-crystal template mechanism. Nature **359**, 710–712 (1992)
7. J.S. Beck, J.C. Vartulli, W.J. Roth, M.E. Leonowicz, C.T. Kresge, K.D. Schmitt, C.T.W. Chu, D.H. Olson, E.W. Sheppard, S.B. McCullen, J.B. Higgins, J.L. Schlenker, A new family of mesoporous molecular sieves prepared with liquid crystal templates. J. Am. Chem. Soc. **114**, 10834–10843 (1992)
8. A.P. Côté, A.I. Benin, N.W. Ockwig, M. O'Keeffe, A.J. Matzger, O.M. Yaghi, Porous, crystalline, covalent organic frameworks. Science **310**, 1166–1170 (2005)
9. M. Dogru, T. Bein, On the road towards electroactive covalent organic frameworks. Chem. Commun. **50**, 5531–5546 (2014)
10. B.J. Smith, W.R. Dichtel, Mechanistic studies of two-dimensional covalent organic frameworks rapidly polymerized from initially homogenous conditions. J. Am. Chem. Soc. **136**, 8783–8789 (2014)
11. Q. Ji, R.C. Lirag, O.S. Miljanic, Kinetically controlled phenomena in dynamic combinatorial libraries. Chem. Soc. Rev. **43**, 1873–1884 (2014)
12. C. Sanchez, C. Boissière, D. Grosso, C. Laberty, L. Nicole, Design, synthesis, and properties of inorganic and hybrid thin films having periodically organized nanoporosity. Chem. Mater. **20**, 682–737 (2008)
13. T.Y. Ma, Synthesis and applications of ordered mesoporous metal phosphonate and sulfonate materials. Doctorial Dissertations, Nankai University (2013)
14. X.Z. Lin, Synthesis and characterization of mesostructured metal organophosphonates. Doctorial Dissertations, Nankai University (2013)
15. J. Livage, Vanadium pentoxide gels. Chem. Mater. **3**, 578–593 (1991)
16. J. Livage, M. Henry, C. Sanchez, Sol-gel chemistry of transition metal oxides. Prog. Solid State Chem. **18**, 259–341 (1988)
17. J. Livage, Palladium-catalysed cross-coupling reactions of ruthenium bis-terpyridyl complexes: strategies for the incorporation and exploitation of boronic acid functionality. New J. Chem. **25**, 1136–1147 (2001)
18. M. Chatry, M. Henry, M. In, C. Sanchez, J. Livage, The role of complexing ligands in the formation of non-aggregated nanoparticles of zirconia. J. Sol-Gel. Sci. Technol. **1**, 233–240 (1993)
19. K. Nakanishi, K. Kanamori, Organic–inorganic hybrid poly(silsesquioxane) monoliths with controlled macro- and mesopores. J. Mater. Chem. **15**, 3776–3786 (2005)
20. P. Innocenzia, B. Lebeau, Organic–inorganic hybrid materials for non-linear optics. J. Mater. Chem. **15**, 3821–3831 (2005)
21. S. Inagaki, S. Guan, Y. Fukushima, T. Ohsuna, O. Terasaki, Novel mesoporous materials with a uniform distribution of organic groups and inorganic oxide in their frameworks. J. Am. Chem. Soc. **121**, 9611–9614 (1999)
22. T. Asefa, M.J. MacLachlan, N. Coombs, G.A. Ozin, Periodic mesoporous organosilicas with organic groups inside the channel walls. Nature **402**, 867–871 (1999)
23. B.J. Melde, B.T. Holland, C.F. Blanford, A. Stein, Mesoporous sieves with unified hybrid inorganic/organic frameworks. Chem. Mater. **11**, 3302–3308 (1999)
24. J. Perles, N. Snejko, M. Iglesias, M. Ángeles, Monge, 3D scandium and yttrium arenedisulfonate MOF materials as highly thermally stable bifunctional heterogeneous catalysts. J. Mater. Chem. **19**, 6504–6511 (2009)
25. H. Furukawa, K.E. Cordova, M. O'Keefe, O.M. Yaghi, The chemistry and applications of metal-organic frameworks. Science **341**, 974–986 (2013)

26. T.Y Ma, Z.Y. Yuan, Metal phosphonate hybrid mesostructures: environmentally friendly multifunctional materials for clean energy and other applications. ChemSusChem **4**, 1407–1419 (2011)
27. Y.P. Zhu, T.Z. Ren, Z.Y. Yuan, Mesoporous non-siliceous inorganic-organic hybrids: a promising platform for designing multifunctional materials. New J. Chem. **38**, 1905–1922 (2014)
28. Y.P. Zhu, T.Y. Ma, Y.L. Liu, T.Z. Ren, Z.Y. Yuan, Metal phosphonate hybrid materials: from densely layered to hierarchically nanoporous structures. Inorg. Chem. Front. **1**, 360–383 (2014)

Chapter 2
History and Classification of Non-Siliceous Hybrid Materials

Abstract The exploration and creation of advanced materials make vital contributions to the development of human civilization. In other words, the human history is the unceasing innovation of materials since the ancient times. Indeed, our living system, Mother Nature, has provided us plenty of amazing presents such as water, food, and pristine tools. Among them, hybrid materials have gradually received our interest owing to the fantastic physicochemical properties constructed from the intimately integrated organic and inorganic units. Noticeably, the exploration concerning the "organic–inorganic hybrid materials" did have a long history. The establishment of a research system started in the middle of last century. With the rapid development of science and technology, the cognition of hybrid materials can indeed reach the molecular level. Furthermore, organic–inorganic hybrids gradually play vital roles in scientific research, industrial production, and even our daily life. Therefore, it is quite necessary to present the history and development of hybrid materials.

Keywords Classification of hybrid materials · Non-siliceous · Metal phosphonates · Metal sulfonates · Metal carboxylates

2.1 Brief History of Hybrid Materials

Organic–inorganic hybrid materials are typically described as the intimate integration of organic and inorganic moieties at the molecular scale and thus fall within the category of nanocomposite material. Indeed, nature provides suitable conditions for the generation of organic–inorganic hybrids such as mollusk shells, crustacean carapaces, and bone [1]. Deliberate efforts to combine the favorable properties of inorganic units and organic moieties in a single composite material represent an old challenge that started at the beginning of the industrial era. However, the first hybrid material made by humanity appeared only very recently at the geologic time scale. Because of the inherited natural availability and the intrinsic properties including adsorption capability, ion-exchange ability, and

© The Author(s) 2015
Y.-P. Zhu and Z.-Y. Yuan, *Mesoporous Organic-Inorganic Non-Siliceous Hybrid Materials*, SpringerBriefs in Molecular Science,
DOI 10.1007/978-3-662-45634-7_2

favorable chemical and physical stabilities, hybrid materials on the basis of the organically modified clays represented an indispensable part of human creation and were gradually employed along the history of artistic, social, industrial, and commercial uses [2, 3]. In America for instance, the ancient Maya site contained an impressive collection of fresco paintings characterized by bright blue and ocher colors, which was known as Maya blue. This pigment was resulted from the introduction of a natural organic dye (blue indigo) into the channels of microfibrous clay (palygorskite) [4]. In China, hybrid clays allowed the production of very thin ceramics thanks to the intercalation of urea inside the interlayer space, facilitating the further delamination which enhanced the resulting plasticity [5].

Organic–inorganic hybrids presented a strong scientific and industrial development over the twentieth century, during which the refined analytical methods and techniques allowed researchers to understand the true natures and structures. The notion of mixing organic and inorganic components has been part of the manufacturing technologies since 1940s [6]. For example, silicones, nanopigments suspended in organic mixtures, and organically templated zeolites provided a diversity of functional hybrid materials that have found application potential in various industrial and scientific research fields. Nonetheless, the concept "hybrid materials" was not proposed at that time. At the end of 1950s, several scientific communities made valuable contributions to the domain of mixed organic–inorganic compounds, which concerned the intercalation of organic units inside the clay and inorganic lamellar compounds [7, 8]. With the establishment of "*chimie douce,*" Livage opened the gates toward a new galaxy of materials in the middle 1980s, namely hybrid materials [9–13], and the concept of "organic–inorganic hybrid nanocomposites" exploded in the 1990s [14, 15]. The period between 1980 and 1995 was particularly fruitful due to the scientific melting pot resulting from the establishment of solgel chemistry (Fig. 2.1). The meeting between material scientists committing to glass and ceramics with chemists mainly working on polymers promoted the tremendous growth in creating a mass of mixed organic–inorganic composites.

Due to the mild conditions involved in the solgel process, solgel-derived siliceous species can be further modified or functionalized with polymers,

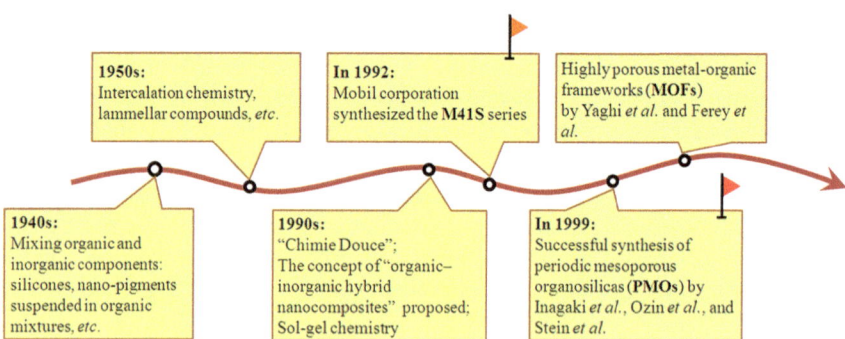

Fig. 2.1 Representative events of the development history for hybrid and porous materials

macromonomers, and numerous organic dyes or biomolecules, leading to the generation of two important materials known as "solgel optics" and biohybrids [16–20]. The synthesis of materials from the polymerization of organosilanes and metal alkoxieds [21] and the design of transition-metal oxide-based composites particularly using nanobuilding block methodologies [22–24] represent major advances in the field as well.

The name of "hybrid materials" was evoked around the 1990s when the input of molecular chemistry was obviously creating a "scientific tsunami" in the domain of nanomaterials science [25]. In the period of 1990–1995, the feasibility of adjustment over the textures, structures, and compositions of hybrids was facilitated with the development of the chemistry of bridged and cubic polysilsesquioxane [26–30]. The 1990s were quite productive with the birth of two meaningful research directions depending on different strategies of hybrid chemistry. The first one concerned the synthesis of periodically organized mesoporous materials obtained through solgel condensation templated via the formation of micellar lyotropic assemblies generated by amphiphilic molecules or polymers. These vital works were pioneered by Inagaki et al., Ozin et al., and Stein et al. in 1999 [31–33]. Thereafter, Inagaki and co-workers successfully prepared an ordered benzene–silica via a surfactant-assisted method, showing a hexagonal array of mesopores with a lattice constant of 52.5 Å, and crystal-like pore walls that exhibit structural periodicity with a spacing of 7.6 Å along the channel direction (Fig. 2.2) [34]. The other domain is related to the very interesting family of hybrids known as metal–organic frameworks (MOFs) that can be categorized as nanoporous and crystalline hybrid coordination polymers. In fact, this kind of porous crystalline materials has a long history, and classical examples include transition metal cyanide compounds, such as Hofmann-type clathrates, Prussian blue-type structures, and Werner complexes, and the diamond-like framework *bis*(adiponitrilo)copper(I)

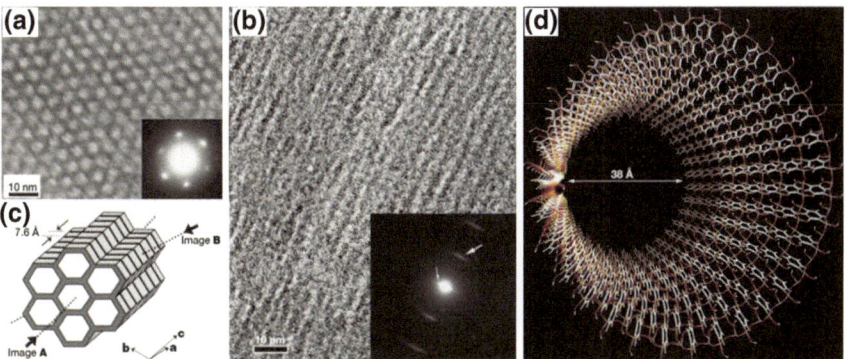

Fig. 2.2 TEM images, electron diffraction patterns, and the resulting structural model of mesoporous benzene–silica (**a**, **b**). Schematic model of mesoporous benzene–silica derived from the results of the TEM images and electron diffraction patterns (**c**). Model showing the pore surface of mesoporous benzene–silica (**d**). Reprinted with permission from Ref. [34]. Copyright 2002, Nature Publishing Group

nitrate [35]. However, there was little interest in such materials until the middle 1990s, when several groups, particularly those of Robson and Yaghi, recognized that rigid, polyfunctional organic molecules could be used to bridge metal cations or clusters into extended arrays with large voids. To produce these robust materials, one could envision constructing the equivalent of a "molecular scaffold" by connecting rigid rod-like organic moieties with inflexible inorganic clusters or single metal centers that act as joints. The size, and more importantly, the chemical environment of the resulting architectures, and void spaces can be precisely controlled by the length and functionalities of the organic units.

Generally speaking, the organic ligating groups involved in MOFs are carboxylates and pyridine-based linkers. As to phosphonic groups, metal phosphonates have been studied as layered inorganic networks, and later evolved into organic–inorganic hybrids by having organic pillars appended off the rigid inorganic layers [36]. Layered structures are predominant for most metals with the organic groups being oriented perpendicularly into the interlamellar region (Fig. 2.3) [37]. Thus the interlayer distance can be easily tuned by changing the pillar groups and it is even possible to exfoliate the layers into film [38]. It should be recognized that the pillars are too crowded and insufficient free space remains in the interlayer region, and no or poor porosity is expected to be present. Several tactics have been adopted to create porosity in the metal phosphonate frameworks, such as substituting phosphonic acid by some non-pillaring groups, extending the geometry of polyphosphonic linkages, and attaching secondary functional groups. Noticeably, organosulfonic acids are considered as relatively poor ligands by coordination chemists and have been used as "non-coordinating" anions in the past synthetic and structural investigation [39]. Until recently, the coordination chemistry of

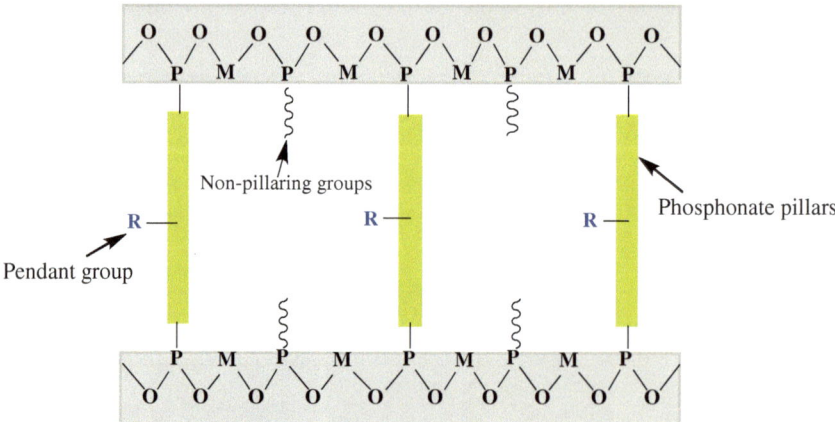

Fig. 2.3 Schematic representation of layered metal phosphonates. To improve the porosity, one way is to insert small non-pillaring "spacer" groups, including metal oxide clusters, phosphoric, phosphors, methylphosphonic acids, and crown ethers, between the phosphonate pillars. Another route is to attach pendant functional groups, such as imino, pyridine, hydroxyl, carboxylic acid, and sulfonic acid, on the organophosphonic linkages

organosulfonate groups has remained relatively unstudied, which can be due to the weak interaction or non-interaction capacity with the majority of transition metal ions or complexes.

Nowadays, the research of hybrid materials has shifted toward much more sophisticated nanocomposites with higher added values. The field of organic–inorganic material has been broadened to a multidisciplinary area, including organometallics, colloids and nanoobjects, soft matter and polymers, coordination polymers such as MOFs, sol/gel, aerosol/aerogel, catalysis and interfaces, porous materials, clays and lamellar compounds, nanocomposites, biomaterials, and bioengineering. Furthermore, a very significant trend is the growing research interest in the rational design of functional hybrids, which extends the field even further. Hybrid materials represent an inexhaustible source of inspiration for us to explore and discover.

2.2 Classification of Non-Siliceous Hybrid Materials

Organic–inorganic hybrids can be defined as nanocomposite materials with intimately linked organic bridging groups and inorganic units. The versatile changes in composition and structure can bring various physicochemical properties that are not the simple sum of the individual contribution of both construction phases. As a result, the nature of the interface and the interactions between the organic and inorganic units can be employed to categorize the hybrids into two main classes [15, 40, 41]. Class I is associated with the hybrid systems that involve no covalent or weak chemical bonding. In this class, only hydrogen bonding, van der Waals or electrostatic forces are usually present. Conversely, Class II hybrid materials show strong chemical interactions between the components, which are formed when the discrete inorganic building blocks are covalently bonded to the organic polymer or inorganic and organic polymers are covalently connected with each other [42, 43]. On the other side, hybrids can also be characterized by the type and size of the organic or the inorganic precursors [15, 40]. Precursors can be two separate monomers or polymers and even covalently linked ones. Because of the mutual insolubility between inorganic and organic components, phase separation will occur. However, homogeneous or single-phased hybrids can be obtained through judiciously choosing bifunctional monomers that contain organic and inorganic components, or by combining both types of components in the phases where one of them is in large excess [44].

The chemical strategies to construct Class II hybrid frameworks are dependent on the relative stability of the interactions between the components and the chemical linkages that connect different components. PMOs represent the typical examples of Class II hybrids. The stable Si–C bonds under hydrolytic conditions allow for the easy incorporation of a large variety of organic bridges in the silica network during the solgel process. Nowadays, the potential of hybrid materials is further strengthened due to the fact that many of them are entering various markets. From the academic and industrial point of view, the increasing impact of hybrid materials science can be summarized following the arborescent representation (Fig. 2.4). New systems should be aimed at high levels of sophistication and miniaturisation, recyclability, eco-friendliness, and cost efficiency.

Recently, the research focus has been turned toward metal phosphonates, sulfonates, and carboxylates. A diversity of organophosphonic, organosulfonic, and organocarboxylic acids and corresponding derivatives (i.e., salts and esters) has been discovered in nature. Judicious design of the organic bridging groups can introduce desirable properties into the hybrid frameworks. The different reactivity of coupling molecules leads to structural diversity and physicochemical peculiarities of the resultant hybrid materials and may provide decisive advantages in the

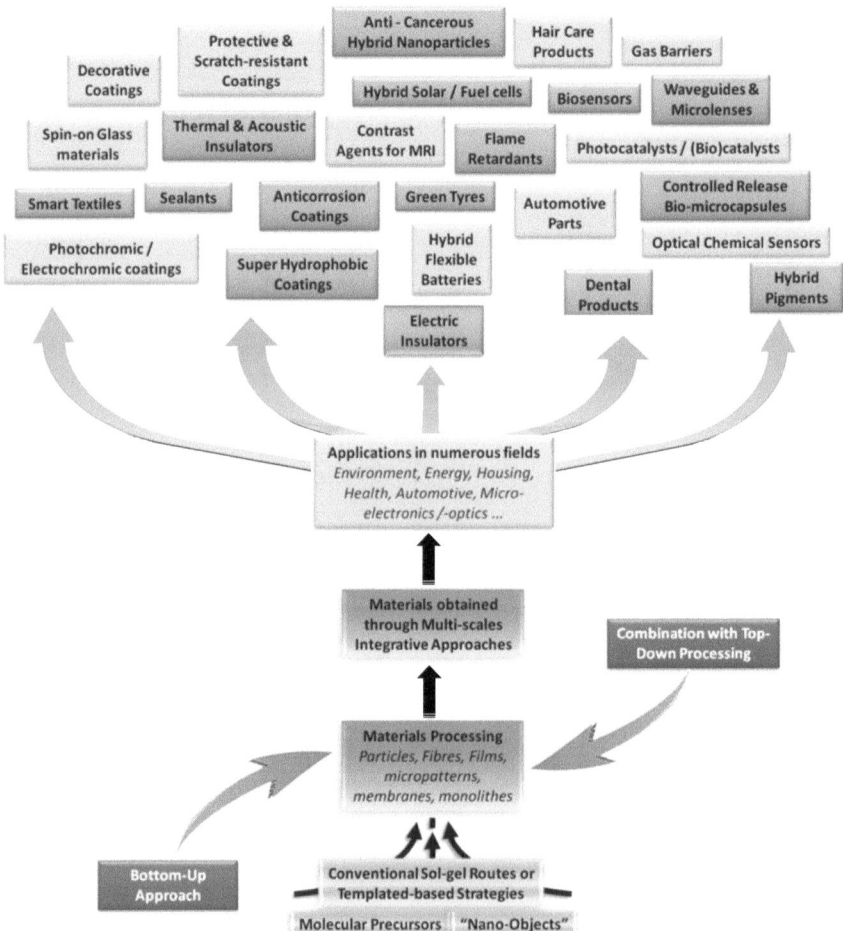

Fig. 2.4 Arborescence representation of hybrid materials. Reprinted with permission from Ref. [11]. Copyright 2011, Royal Society of Chemistry

synthesis of homogeneous hybrids. The homogeneous and efficient incorporation of organic functional groups into the framework of the materials can be realized, allowing for uniform physicochemical properties from the external surface to the internal skeleton.

Metal phosphonate chemistry is originated from the corresponding inorganic phosphate counterparts. In the beginning, gels were refluxed in strong H_3PO_4 and crystallized into what later came to be known as α-zirconium phosphate (α-ZrP) [45]. This compound has a clay-like structure in which the ZrO_6 octahedra are sandwiched between layers of phosphate tetrahedral. Dines et al. first conceived of producing porous materials by cross-linking the α-zirconium phosphate-type layers using diphosphonic acids, $H_2O_3P\text{-}R\text{-}PO_3H_2$, where R may be an alkyl or aryl group [46]. The strategy was to choose the cross-linking groups that are large and then space them such that different size of pores would result. However, the area subtended by a phosphate group on the α-ZrP layer was 24 Å^2. Given the fact that an alkyl or aryl group spaced every 5.3 Å apart on the layer occupies most of the area between pillars, there should be no microporosity. To overcome this restriction, the Dines group used phosphorous acid as a spacer group, together with biphenyl as the R group. The idea was to space the biphenyl pillars two or three positions apart, thus creating microporosity. Clearfield and co-workers contributed a lot to the intimately relevant work concerning diphosphonate derivatives [47], which contains 2D sheets of ZrO_6 octahedra sandwiched between phosphate (or phosphonate) layers, which were akin to many other inorganic clays. Porous zirconium diphosphonates are synthesized by combining both a rigid diphosphonate (such as biphenylene bis(phosphonate)) and phosphite (HPO_3^-) or phosphate, where the average pore size could be adjusted by varying the ratio of acids used in the synthesis [48]. Figure 2.5 illustrates a possible structure for nanoporous zirconium diphenylenebis(phosphonate)/phosphate.

One important advantage of this approach is that materials with pores in the range of micropores and mesopores may be obtained. That is to say, an advantage

Fig. 2.5 An idealized structure of a porous zirconium phosphonate. ZrO_6 octahedra are shown in gray, PO_3C tetrahedra in light gray, and carbon atoms as light gray spheres

of the relatively small crystallite size and flat-plate morphology leads to the resulting materials with both nano- and mesoporosity. There has been some work to add functionality to these materials through the phosphonate used in the synthesis. Vermuelen and Thompson used viologens, rigid organic molecules of the formula $(H_2O_3PCH_2CH_2\text{-bipyridinium-}CH_2CH_2PO_3H_2)$, to bridge zirconium phosphonate layers [49]. These molecules have potentially useful photoinduced charge separation properties, and the presence of two positive charges on the amines in the bridge leads to the incorporation of anions into the framework for charge balance. Although the optical properties of the resulting phase were intriguing, the structure was dense and differed significantly from that normally encountered in zirconium phosphonate chemistry [50]. However, substitution of a fraction of the viologen diphosphonate with phosphoric acid resulted in porous materials [51] with ordered channels [52]. Owing to the tunable pore sizes, easy functionalization, and moderate thermal stability, metal phosphates/phosphonates become one of the most promising hybrid materials.

The metal-ligand coordinate bonds between donor and acceptor molecules have been exploited to generate infinite framework structures. For the synthesis of porous materials, networks are often envisioned where rigid organic molecules and metal atoms or clusters replace bonds and atoms in classical inorganic structures. Copper tetra(4-cyanophenyl) methane contains two types of tetrahedral nodes; Cu^+ coordinated to 4-cyano groups and the methane center in the organic ligand [53]. Together, these define a tetrahedral network with the same topology as the diamond structure with phenyl groups replacing carbon–carbon bonds. Yaghi has been instrumental in applying this methodology to carboxylate systems. But the inability of these solids to maintain permanent porosity and avoid structural rearrangements upon guest removal or guest exchange has been an obvious shortcoming. Carboxylate-based metal–organic frameworks that exhibit permanent porosity have now been prepared [54–56]. The first such solid was MOF-5, which consists of Zn^{2+} and 1,4-benzenedicarboxylate and has a microporous volume larger than any known zeolite [54]. There are now annually hundreds of papers describing MOFs, most of which possess highly complex structures. The Férey group has prepared many open framework carboxylates. The reactions carried out hydrothermally at 220 °C, and HF additions could achieve sufficient crystallinity for structure solutions from powders of new Cr(III) compounds [57]. Subsequently, Férey et al. [58, 59] prepared two additional chromium terephthalates containing pores of 25–29 Å and a surface area of about 3,100 m^2 g^{-1} (Langmuir).

In addition to metal phosphonates and carboxylates, metal sulfonates represent another significant member of non-siliceous hybrid family. The coordination chemistry of the sulfonic groups has been less thoroughly investigated than that of carboxulic and phosphonic ones, probably due to the weakly coordinating behavior usually attributed to the sulfonate ligand. Nevertheless, the sulfonic group has a wide variety of possible coordinative modes and has been reported to form several types of layered or pillared layered compounds [60–62] with silver, alkali and alkaline earth metal ions, as well as transition metals and lanthanide(III) ions.

Noticeably, the majority of transition metal aqua complexes with sulfonate counter anions show that the sulfonate group cannot readily displace water from the coordination sphere of the metal ion [63]. However, a stable solid can be yielded when suitably soft metal cations are employed, including alkali ions, larger alkaline earth ions, and silver(I). The common feature of these ions is that none of them have stringent preferences with respect to coordination number or geometry.

As compared with the carbon atoms in the carboxylic groups, the central atoms of the sulfonic and phosphonic acids are able to accommodate more than eight electrons in the outer electron shell, which accounts for a greater bonding flexibility in these groups [64–66]. Nonetheless, the proton of the sulfonic group is more easily dissociable than the proton of the carboxylic and phosphonic functional groups and thus in this respect the sulfonic acids are stronger than their phosphonic and carboxylic counterparts. On the other hand, sulfonic acids can fully deprotonate at very low pK_a, causing serious problem with their stability. This implies the relatively weak coordination interactions between the sulfonate anions and metal cations, which make the frameworks insufficiently robust to sustain permanent porosity [39, 64]. In comparison with their sulfonate and carboxylate analogs, metal phosphonates exhibit much higher thermal and chemical stability due to the strong affinity of organophosphonic linkers to metal ions, making them promising in the fields of energy conversion, adsorption/separation, catalysis, biotechnology, and so forth [67, 68]. As to phosphonates and sulfonates, the coordination chemistry is quite similar, though the corresponding coordination is less predictable owing to more possible ligating modes and three probable states of protonation relative to the carboxylate bridging groups. Typical coordination modes between organic linkages and metal ions are illustrated in Fig. 2.6. For a single phosphonate/sulfonate group, each oxygen atom possesses the capacity to bridge more than one metal center. While formally a single phosphonate/sulfonate oxygen atom can bind to three metal centers, more typically, the oxygen

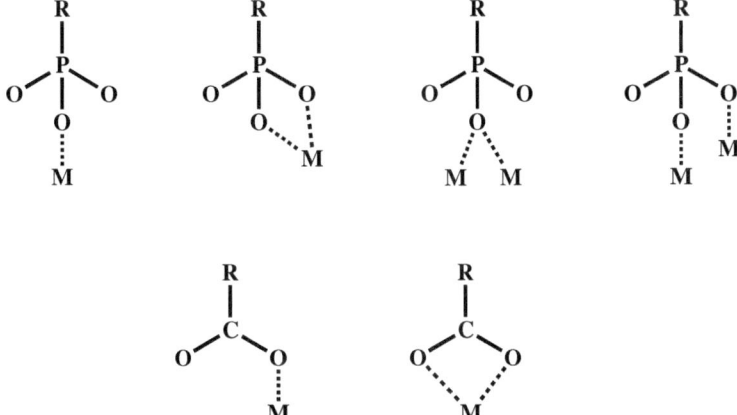

Fig. 2.6 Typical coordination modes of phosphonates and carboxylates

atoms of a phosphonate/sulfonate linker will bridge two metal ions maximum. The large bridging numbers are unattainable with other commonly employed classes of ligands such as carboxylates [69]. Noticeably, whereas the ligating directionality of a carboxylate group is confined to a plane, the spherical ligating shape of the electron density that encompasses a phosphonate/sulfonate group allows metal coordination to an additional dimension, which serves to further increase the connectivity of the network and favor the formation of a robust structure. This coordinative flexibility in terms of bridging modes, combined with the roughly spherical shape of the PO_3/SO_3 unit, has led us to draw the analogy between a phosphonate/sulfonate group and a "Ball of Velcro." It should be noted that the chelation capacity of the organic linkages to metal ions usually follows the sequence of phosphonates > carboxylates > sulfonates [39, 64]. Therefore, the predisposition of simple metal phosphonates to a dense layered motif makes forming high surface area materials a challenge [70]. Numerous methodologies have thus been developed to incorporate considerable mesoporosity in the non-siliceous hybrid framework.

2.3 General Strategies of Incorporating Organic Groups

2.3.1 Surface Functionalization

Surface modification can have a significant influence on the materials behavior at the nanoscale and can lead to nano-/mesostructures with novel properties. Post-synthetic functionalization or grafting refers to the subsequent modification of the inner surfaces of mesostructured inorganic phases with organic groups (Fig. 2.7).

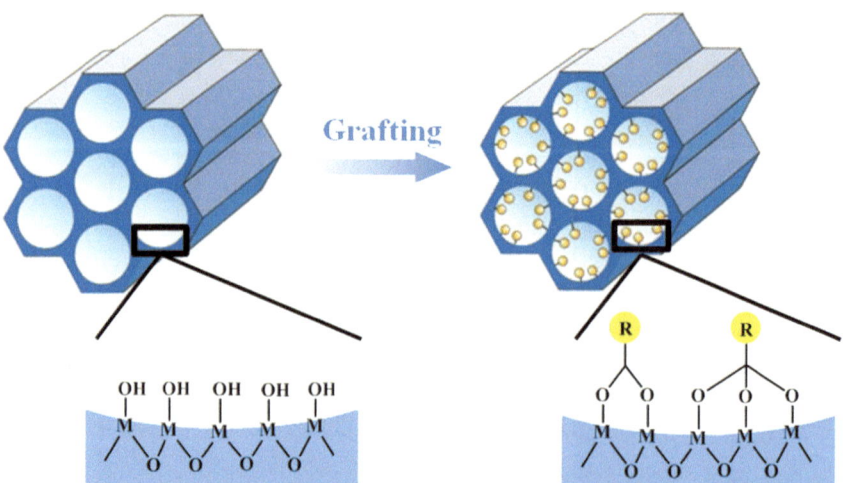

Fig. 2.7 Surface modification of mesoporous inorganic frameworks, R represents organic functional group

The absence of homocondensation between organic functional groups and the easy formation of organic–inorganic layers allows the straightforward and reproducible formation of robust monolayers on an extremely wide range of substrates. On one hand, the original physicochemical properties of the bulk materials can be preserved. On the other hand, diverse novel functional groups with special properties including bioactivity, electronic conductivity, and photochemical properties can be incorporated, showing the capacity to be further modified as well.

Mallouk et al. proposed to employ bisphosphonic acids for the formation of metal phosphonate multilayers on surfaces [71]. Later on, Guerrero et al. reported the anchoring of phosphonate and phosphinate coupling molecules on titania particles [72], with the use of six organophosphorous compounds: phenylphosphonic and diphenylphosphonic acids, their ethyl esters, and their trimethylsilyl esters. In the case of organophosphorus coupling molecules, reaction with the surface involves not only the condensation with surface hydroxyl groups but also the coordination of the phosphoryl on Lewis acid sites, and the cleavage of the M–O–M bonds depending on the anchoring conditions. The hydrolytic stability of organic monolayers supported on metal oxides was also investigated [73]. It was found that the monolayers of $C_{18}H_{37}P(O)(OH)_2$ demonstrated a better hydrolytic stability than other octadecyl organosilane modifiers. The high stability of these phosphonate monolayers is explained by the strong specific interactions of the phosphonic acid group with the surfaces of metal oxides. On the basis of the above-mentioned literature reports, the feasibility of grafting phosphonic acids onto metal oxides is fully confirmed. Soler-Illia et al. prepared organic modified transition metal oxide mesoporous thin films and xerogels by using dihexadecyl phosphate (DHDP), monododecyl phosphate (MDP), and phenyl phosphate (PPA) [74]. Dramatic differences were observed for the incorporation of organophosphonates in mesoporous versus non-mesoporous solids, demonstrating that the organic functions were incorporated inside the pore system. Incorporation behaviors were also observed depending on the mesostructure; cubic 3D mesostructures are more accessible than their 2D hexagonal counterparts [75]. Furthermore, the functionalized pores were found to be further accessible to other molecules (solvent and fluorescent probes) or ions (i.e., Hg^{2+}), opening the way for sensor or sorption applications.

Besides the monophosphonic acids mentioned above, Yuan and co-workers reported the use of a series of amine-based organophosphonic acids and their salts as organophosphorus coupling molecules in the one-step synthesis and the application exploration of oxide–phosphonates and metal organophosphonate hybrid materials with mesopores and hierarchical meso-/macroporous architectures [76, 77]. Claw molecules of ethylene diamine tetra(methylene phosphonic acid) (EDTMP) and diethylene triamine penta(methylene phosphonic acid) (DTPMP) were anchored to the titania network homogeneously. The synthesized titania–phosphonate hybrids showed irregular mesoporosity formed by the assembly of nanoparticles in a crystalline anatase phase. The synthesis process is quite simple in comparison with the previously reported two-step solgel processing involving first the formation of P–O–M bonds by non-hydrolytic condensation of a metal alkoxide with a phosphonic acid and then the formation of the M–O–M bonds of the metal oxide network by hydrolysis/condensation of the remaining alkoxide group. The burdensome work to remove the residual organic solvent was not needed.

Organic "capping" agents containing terminal carboxylic groups have been widely used to prepare quantum dots or nanocrystals [78–81]. The capping agent limits the size of the nanoparticles by preventing further particle growth and agglomeration during synthesis and can also be useful in controlling the particle reactivity, imparting solubility or packing characteristics, and protecting both nanoparticle and its environment from destructive interactions. If different functional groups are anchored on the carboxylate linkages, variation of tremendous physicochemical properties can be realized. To date, reports regarding carboxylated mesoporous material are rare because the relatively low stability of C–O–M may prohibit their practical applications.

Anchoring of the functional groups is driven by condensation or by complexation. Correspondingly, grafting can be covalent (practically irreversible) or coordinative (partially reversible). Stronger grafting groups are needed for sensing and catalysis, while more labile functions are meaningful in the quest for controlled delivery or reversible signaling as well. The grafting strength will be also important for the even incorporation of the R function along the pore systems. Indeed, there are two key factors that control the homogeneous incorporation of organic functions: the accessibility of the pore systems; and the reactivity of the organic functional molecules toward the pore surface. The first factor will essentially depend on the possibility of pore interconnection, and the symmetry and orientation of the pore mesostructure. Noticeably, pore blocking can occur during the post-functionalization process.

2.3.2 Direct Synthesis

An alternative method to synthesize organically functionalized mesoporous hybrid is the cocondensation method (one-pot synthesis). It is possible to prepare mesostructured hybrid phases by the cocondensation of metallic precursors and organic functional groups in the presence or absence of structure-directing agents, leading to materials with organic residues anchored covalently and homogeneously in the pore walls (Fig. 2.8) [64, 65, 82]. By using structure-directing agents known from the synthesis of pure mesoporous silica phases (e.g., MCM or SBA silica phases), organically modified silicas can be prepared in such a way that the organic functionalities project into the pores. These will be elaborately discussed in Chap. 3.

Since the organic functionalities are direct components of the mesostructured hybrid matrixes, pore blocking is not a problem in the cocondensation method. Furthermore, the organic linkers are generally more homogeneously distributed than in materials synthesized involving the grafting process. The tendency toward homocondensation reactions, which is caused by the different hydrolysis and

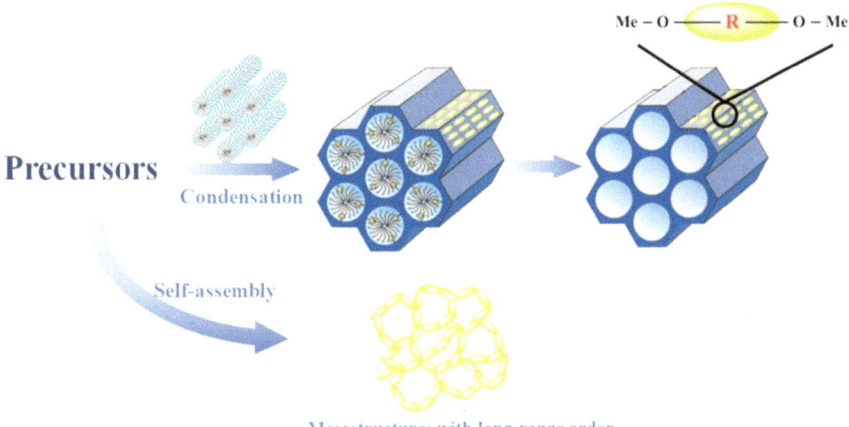

Fig. 2.8 General synthetic pathway to mesoporous non-siliceous hybrid materials

condensation rates of the structurally different precursors, is a constant problem in cocondensation because the homogeneous distribution of different organic functionalities in the framework cannot be guaranteed. Furthermore, an increase in loading of the incorporated organic groups and the complexity of organic linking groups can lead to a deterioration of the mesoporosity including the pore diameter, pore volume, and specific surface areas. Purely methodological disadvantage that is associated with the cocondensation method is that care must be taken not to destroy the organic functionality during the removal of the surfactant. This is why commonly extractive methods are favorable, while calcination is not suitable in most cases.

References

1. S. Mann, *Biomimetic Materials Chemistry* (Wiley-VCH, Weinheim, 1997)
2. E. Ruiz-Hitzky, P. Aranda, M. Darder, G. Rytwo, Hybrid materials based on clays for environmental and biomedical applications. J. Mater. Chem. **20**, 9306–9321 (2010)
3. F. Bergaya, G. Lagaly, *Handbook of Clay Science*, Elsevier Science, 2013
4. H. Berke, The invention of blue and purple pigments in ancient times. Chem. Soc. Rev. **36**, 15–30 (2007)
5. A. Weiss, A secret of Chinese porcelain manufacture. Angew. Chem. Int. Ed. **2**, 697–703 (1963)
6. B. Arkles, Commercial applications of sol-gel-derived hybrid materials. MRS Bull. **26**, 402–407 (2001)
7. M. Darder, P. Aranda, E. Ruiz-Hitzky, Bionanocomposites: a new concept of ecological, bioinspired, and functional hybrid materials. Adv. Mater. **19**, 1309–1319 (2007)
8. E. Ruiz-Hitzky, P. Aranda, M. Darder, G. Rytwo, Hybrid materials based on clays for environmental and biomedical applications. J. Mater. Chem. **20**, 9306–9321 (2010)
9. C. Laberty-Robert, K. Vallé, F. Pereira, C. Sanchez, Design and properties of functional hybrid organic-inorganic membranes for fuel cells. Chem. Soc. Rev. **40**, 961–1005 (2011)

10. D. Avnir, D. Levy, R. Reisfeld, The nature of the silica cage as reflected by spectral changes and enhanced photostability of trapped Rhodamine 6G. J. Phys. Chem. **88**, 5956–5959 (1984)

11. C. Sanchez, P. Belleville, M. Popalld, L. Nicole, Applications of advanced hybrid organic-inorganic nanomaterials: from laboratory to market. Chem. Soc. Rev. **40**, 696–753 (2011)

12. J. Livage, M. Henry, C. Sanchez, Sol-gel chemistry of transition metal oxides. Prog. Solid State Chem. **18**, 259–341 (1988)

13. C. Sanchez, J. Livage, M. Henry, F. Babonneau, Chemical modification of alkoxide precursors. J. Non-Cryst. Solids **100**, 65–76 (1988)

14. C. Sanchez, F. Ribot, Proceedings of the first European workshop on hybrid organic-inorganic materials. New J. Chem. **18**, 987–988 (1993)

15. C. Sanchez, F. Ribot, New J. Chem. **18**, 1007–1047 (1994)

16. F. Surivet, T.M. Lam, J.P. Pascault, Q.T. Pham, Organic-inorganic hybrid materials. 1. Hydrolysis and condensation mechanisms involved in alkoxysilane-terminated macromonomers. Macromolecules **25**, 4309–4320 (1992)

17. F. Surivet, T.M. Lam, J.P. Pascault, C. Mai, Organic-inorganic hybrid materials. 2. Compared structures of polydimethylsiloxane and hydrogenated polybutadiene based ceramers. Macromolecules **25**, 5742–5751 (1992)

18. T. Fournier, I. Salabert, T.H. Tran-Thi, H. Ali, J. Van-Lier, C. Sanchez, Charge transfer dynamics of donor-acceptor systems in solutions and sol-gel matrices. J. Sol-Gel. Sci. Technol. **2**, 737–740 (1993)

19. C. Sanchez, B. Lebeau, F. Chaput, J.P. Boilot, Optical properties of functional hybrid organic-inorganic nanocomposites. Adv. Mater. **15**, 1969–1994 (2003)

20. D. Avnir, T. Coradin, O. Lev, J. Livage, Recent bio-applications of sol-gel materials. J. Mater. Chem. **16**, 1013–1030 (2006)

21. P. Audebert, C. Sanchez, Modified electrodes from hydrophobic alkoxide silica gels-Insertion of electroactive compounds and glucose oxidase. J. Sol-Gel. Sci. Technol. **2**, 809–812 (1993)

22. G. Guerrero, J.G. Alauzun, M. Granier, D. Laurencin, P.H. Mutin, Phosphonate coupling molecules for the control of surface/interface properties and the synthesis of nanomaterials. Dalton Trans. **42**, 12569–12585 (2013)

23. C. Sanchez, G.J.A.A. Soler-Illia, F. Ribot, T. Lalot, C.R. Mayer, V. Cabuil, Designed hybrid organic-inorganic nanocomposites from functional nanobuilding blocks. Chem. Mater. **13**, 3061–3083 (2001)

24. L. Rozes, C. Sanchez, Titanium oxo-clusters: precursors for a Lego-like construction of nanostructured hybrid materials. Chem. Soc. Rev. **40**, 1006–1030 (2011)

25. L. Nicole, C. Laberty-Robert, L. Rozes, C. Sanchez, Hybrid materials science: a promised land for the integrative design of multifunctional materials. Nanoscale **6**, 6267–6292 (2014)

26. K.J. Shea, D.A. Loy, Bridged polysilsesquioxanes. Molecular-engineered hybrid organic-inorganic materials. Chem. Mater. **13**, 3306–3319 (2001)

27. K.J. Shea, J. Moreau, D.A. Loy, R.J.P. Corriu, B. Boury, in *Functional Hybrid Materials*, ed. by P. Gomez-Romero, C. Sanchez (WILEY-VCH Verlag GmbH & Co. KGaA, Weinheim, 2004), pp. 50–85

28. G. Cerveau, R.J.P. Corriu, Some recent developments of polysilsesquioxanes chemistry for material science. Coord. Chem. Rev. **178**, 1051–1071 (1998)

29. R.M. Laine, Nanobuilding blocks based on the $[OSiO_{1.5}]x$ ($x = 6, 8, 10$) octasilsesquioxanes. J. Mater. Chem. **15**, 3725–3744 (2005)

30. R. Duchateau, Incompletely condensed silsesquioxanes: versatile tools in developing silica-supported olefin polymerization catalysts. Chem. Rev. **102**, 3525–3542 (2002)

31. S. Inagaki, S. Guan, Y. Fukushima, T. Ohsuna, O. Terasaki, Novel mesoporous materials with a uniform distribution of organic groups and inorganic oxide in their frameworks. J. Am. Chem. Soc. **121**, 9611–9614 (1999)

32. T. Asefa, M.J. MacLachlan, N. Coombs, G.A. Ozin, Periodic mesoporous organosilicas with organic groups inside the channel walls. Nature **402**, 867–871 (1999)

33. B.J. Melde, B.T. Holland, C.F. Blanford, A. Stein, Mesoporous sieves with unified hybrid inorganic/organic frameworks. Chem. Mater. **11**, 3302–3308 (1999)

34. S. Inagaki, S. Guan, T. Ohsuna, O. Terasaki, S. Inagaki, S. Guan, T. Ohsuna, O. Terasaki, An ordered mesoporous organosilica hybrid material with a crystal-like wall structure. Nature **416**, 304–307 (2002)

35. Y. Kinoshita, I. Matsubara, T. Higuchi, Y. Saito, The crystal structure of bis(adiponitrilo)copper (I) nitrate. Bull. Chem. Soc. Jpn **32**, 1221–1226 (1959)

36. A. Clearfield, C.V.K. Sharma, B.L. Zhang, Crystal engineered supramolecular metal phosphonates: crown ethers and iminodiacetates. Chem. Mater. **13**, 3099–3112 (2001)

37. R.C. Finn, R. Lam, J.E. Greedan, J. Zubieta, Solid-state coordination chemistry: structural influences of copper-phenanthroline subunits on oxovanadium organophosphonate phases. Hydrothermal synthesis and structural characterization of the two-dimensional materials [Cu(phen) (VO)(O$_3$PCH$_2$PO$_3$)(H$_2$O)], [{Cu(phen)}$_{(2)}$(V$_2$O$_5$) (O$_3$PCH$_2$CH$_2$PO$_3$)], and [{Cu(phen)}$_{(2)}$(V$_3$O$_5$)(O$_3$PCH$_2$CH$_2$CH$_2$PO$_3$)$_{(2)}$(H$_2$O)] and of the three-dimensional phase [{Cu(phen)}$_{(2)}$(V$_3$O$_5$) (O$_3$PCH$_2$PO$_3$)$_{(2)}$(H$_2$O)]. Inorg. Chem. **40**, 3745–3754 (2001)

38. D.B. Mitzi, Thin-film deposition of organic-inorganic hybrid materials. Chem. Mater. **13**, 3283–3298 (2001)

39. G.K.H. Shimizu, R. Vaidhyanathan, J.M. Taylor, Phosphonate and sulfonate metal organic frameworks. Chem. Soc. Rev. **38**, 1430–1449 (2009)

40. P. Judeinstein, J. Rault, B. Alonso, C. Sanchez, Macroscopic–microscopic mechanical relaxation behavior of hybrid organic–inorganic materials. J. Polymer Sci. Part B Polymer Phys. **39**, 645–650 (2001)

41. R.J.P. Corriu, Chimie douce: wide perspectives for molecular chemistry. A challenge for chemists: control of the organisation of matter. New J. Chem. **25**, 2–2 (2001)

42. T.Z. Ren, Z.Y. Yuan, L.B. Su, Thermally stable macroporous zirconium phosphates with supermicroporous walls: a self-formation phenomenon of hierarchy. Chem. Commun. 2730–2731 (2004)

43. E.G. Vrieling, T.P.M. Beelen, R.A. van Santen, W.W.C. Gieskes, Mesophases of (bio)polymer-silica particles inspire a model for silica biomineralization in diatoms. Angew. Chem. Int. Ed. **41**, 1543–1546 (2002)

44. C. Sanchez, L. Rozes, F. Ribot, C. Laberty-Robert, D. Grosso, C. Sassoye, C. Boissiere, L. Nicole, "Chimie douce": a land of opportunities for the designed construction of functional inorganic and hybrid organic-inorganic nanomaterials. C. R. Chim. **13**, 3–39 (2010)

45. A. Clearfield, Z. Wang, P. Bellinghausen, Highly porous zirconium aryldiphosphonates and their conversion to strong bronsted acids. J. Solid State Chem. **167**, 376–385 (2002)

46. M.D. Dines, P.M. DiGiacomo, K.P. Callahan, P.C. Griffith, R.H. Lane, R.E. Cooksey, in *Chemically Modified Surface in Catalysis and Electrocatalysis*, ed. by J. S. Miller, Chap 12. ACS Symposium Series 192. American Chemical Society, Washington, DC (1982)

47. A. Clearfield, Unconventional metal organic frameworks: porous cross-linked phosphonates. Dalton Trans. **28**(44), 6089–6102 (2008)

48. A. Clearfield, Organically pillared micro- and mesoporous materials. Chem. Mater. **10**, 2801–2810 (1998)

49. L.A. Vermeulen, M.E. Thompson, Stable photoinduced charge separation in layered viologen compounds. Nature **358**, 656–658 (1992)

50. D.M. Poojary, L.A. Vermeulen, E. Vicenzi, A. Clearfield, M.E. Thompson, Structure of a novel layered zirconium diphosphonate compound: Zr$_2$(O$_3$PCH$_2$CH$_2$-viologen-CH$_2$CH$_2$PO3) F$_6$·2H$_2$O. Chem. Mater. **6**, 1845–1849 (1994)

51. L.A. Vermeulen, M.E. Thompson, Synthesis and photochemical properties of porous zirconium viologen phosphonate compounds. Chem. Mater. **6**, 77–81 (1994)

52. H. Byrd, A. Clearfield, D. Poojary, K.P. Reis, M.E. Thompson, Crystal structure of a porous zirconium phosphate/phosphonate compound and photocatalytic hydrogen production from related materials. Chem. Mater. **8**, 2239–2246 (1996)

53. M. O'Keeffe, M. Eddaoudi, H. Li, T. Reineke, O.M. Yaghi, Frameworks for extended solids: geometrical design principles. J. Solid State Chem. **152**, 3–20 (2000)

54. H. Li, M. Eddaoudi, M. O'Keeffe, O.M. Yaghi, Design and synthesis of an exceptionally stable and highly porous metal-organic framework. Nature **402**, 276–279 (1999)

55. B. Chen, M. Eddaoudi, S.T. Hyde, M. O'Keeffe, O.M. Yaghi, Interwoven metal-organic framework on a periodic minimal surface with extra-large pores. Science **291**, 1021–1023 (2001)
56. M. Eddaoudi, J. Kim, N. Rosi, D. Vodak, J. Wachter, M. O'Keefe, O.M. Yaghi, Systematic design of pore size and functionality in isorecticular MOFs and their application in methane storage. Science **295**, 469–472 (2002)
57. F. Millange, C. Serra, G. Férey, Synthesis, structure determination and properties of MIL-53as and MIL-53ht: the first Cr^{III} hybrid inorganic-organic microporous solids: $Cr^{-III}(OH)$ center·$\{O_2C-C_6H_4-CO_2\}$·$\{HO_2C-C_6H_4-CO_2H\}_x$. Chem. Commun. 822–823 (2002)
58. K. Barthelet, J. Marrot, D. Riou, G. Ferey, A breathing hybrid organic-inorganic solid with very large pores and high magnetic characteristics. Angew. Chem. Int. Ed. **41**, 281–284 (2002)
59. G. Ferey, C. Mellot-Draznieks, C. Serre, F. Millange, J. Dutour, S. Suable, I. Margiolaki, A chromium terephthalate-based solid with unusually large pore volumes and surface area. Science **309**, 2040–2042 (2005)
60. D.J. Hoffart, S.A. Dalrymple, G.K.H. Shimizu, Structural constraints in the design of silver sulfonate coordination networks: three new polysulfonate open frameworks. Inorg. Chem. **44**, 8868–8875 (2005)
61. Q. Li, X. Liu, M.L. Fu, C.G. Guo, Two luminescent alkali-silver heterometallic sulfonates. Inorg. Chim. Acta **395**, 2147–2153 (2006)
62. Z.M. Sun, J.G. Mao, Y.Q. Sun, H.Y. Zeng, A. Clearfield, Synthesis, characterization, and crystal structures of three new divalent metal carboxylate-sulfonates with a layered and one-dimensional structure. Inorg. Chem. **43**, 336–341 (2004)
63. A.P. Côté, G.K.H. Shimizu, The supramolecular chemistry of the sulfonate group in extended solids. Coord. Chem. Rev. **245**, 49–64 (2003)
64. Y.P. Zhu, T.Z. Ren, Z.Y. Yuan, Mesoporous non-siliceous inorganic-organic hybrids: a promising platform for designing multifunctional materials. New J. Chem. **38**, 1905–1922 (2014)
65. Y.P. Zhu, T.Y. Ma, Y.L. Liu, T.Z. Ren, Z.Y. Yuan, Metal phosphonate hybrid materials: from densely layered to hierarchically nanoporous structures. Inorg. Chem. Front. **1**, 360–383 (2014)
66. V. Videnova-Adrabinska, Coord. Chem. Rev. **251**, 1987–2016 (2007)
67. A. Clearfield, Z. Wang, Organically pillared microporous zirconium phosphonates. J. Chem. Soc. Dalton Trans. 2937–2947 (2002)
68. M. Pramanik, A. Bhaumik, Self-assembled hybrid tinphosphonate nanoparticles with bimodal porosity: an insight towards the efficient and selective catalytic process for the synthesis of bioactive 1,4-dihydropyridines under solvent-free conditions. J. Mater. Chem. A **1**, 11210–11220 (2013)
69. G.K.H. Shimizu, G.D. Enright, C.I. Ratcliffe, K.F. Preston, J.L. Reid, J.A. Ripmeester, A layered silver sulfonate incorporating nine-coordinate Ag-I in a hexagonal grid. Chem. Commun. 1485–1486 (1999)
70. K. Maeda, Metal phosphonate open-framework materials. Micropor. Mesopor. Mater. **73**, 47–55 (2004)
71. H. Lee, L.J. Kepley, H.G. Hong, T.E. Mallouk, Phase transformations in mesostructured silica/surfactant composites: mechanisms for change and applications to materials synthesis. J. Am. Chem. Soc. **110**, 618–620 (1988)
72. G. Guerrero, P.H. Mutin, A. Vioux, Anchoring of phosphonate and phosphinate coupling molecules on titania particles. Chem. Mater. **13**, 4367–4373 (2001)
73. S. Marcinko, A.Y. Fadeev, Hydrolytic stability of organic monolayers supported on TiO_2 and ZrO_2. Langmuir **20**, 2270–2273 (2004)
74. P.C. Angelomé, G.J.A.A. Soler-Illia, Organically modified transition-metal oxide mesoporous thin films and xerogels. Chem. Mater. **17**, 322–331 (2005)
75. P.C. Angelomé, G.J.A.A. Soler-Illia, Ordered mesoporous hybrid thin films with double organic functionality and mixed oxide framework. J. Mater. Chem. **15**, 3903–3912 (2005)
76. X.J. Zhang, T.Y. Ma, Z.Y. Yuan, Titania-phosphonate hybrid porous materials: preparation, photocatalytic activity and heavy metal ion adsorption. J. Mater. Chem. **18**, 2003–2010 (2008)

77. T.Y. Ma, X.J. Zhang, Z.Y. Yuan, High selectivity for metal ion adsorption: from mesoporous phosphonated titanias to meso-/macroporous titanium phosphonates. J. Mater. Sci. **44**, 6775–6785 (2009)
78. G.H.T. Au, W.Y. Shih, W.H. Shih, High-conjugation-efficiency aqueous CdSe quantum dots. Analyst **138**, 7316–7325 (2013)
79. Z. Hu, M. Ahrén, L. Selegård, C. Skoglund, F. Söderlind, Maria Engström, X. Zhang, K. Uvdal, Highly water-dispersible surface-modified Gd_2O_3 nanoparticles for potential fuel-modal bioimaging. Chem. Eur. J. **19**, 12658–12667 (2013)
80. A. Hassinen, R. Gomes, K.D. Nolf, Q. Zhao, A. Vantomme, J.C. Martins, Z. Hens, Surface chemistry of CdTe quantum dots synthesized in mixtures of phosphonic acids and amines: formation of a mixed ligand shell. J. Phys. Chem. C **117**, 13936–13943 (2013)
81. M.H. Zarghami, Y. Liu, M. Gibbs, E. Gebremichael, C. Webster, M. Law, p-Type PbSe and PbS quantum dot solids prepared with short-chain acids and diacids. ACS Nano **4**, 2475–2485 (2010)
82. T.Y Ma, Z.Y. Yuan, Metal phosphonate hybrid mesostructures: environmentally friendly multifunctional materials for clean energy and other applications. ChemSusChem **4**, 1407–1419 (2011)

Chapter 3
Strategies to Incorporate Mesoporosity

Abstract The development of well-structured mesoporous materials with high surface areas, controllable structures, and the systematic tailoring of pore architecture provides advances in a variety of fields such as adsorption, separation, catalysis, storage, drug delivery, and biosensing. The structural capabilities at the scale of a few nanometers can meet the demands of the applications emerging in large molecules involved in processes. Therefore, scientific researchers across the world have extensively been focused on the exploration and development of mesoporous materials. The introduction of well-defined mesostructures into the ultimate materials seems easy since the key factors are widely known, such as surfactant template and its concentration, temperature, media, precursors, and so forth. In fact, samples synthesized under "similar conditions" present distinctively distinguishing properties. This is reasonable because the complex synthesis systems vary with the changing of microenvironment, and a complicated combination of simple factors will offer great opportunities in creating different porous architectures. The target of different synthesis methodologies is to obtain controllable mesoporosity, thereby fitting the qualifications of a particular application. The synthesis of the porous organic–inorganic hybrid materials is somewhat different from the inorganic porous materials, but their synthesis processes are still comparable. As compared to conventional syntheses of mesoporous siliceous and carbonaceous materials, the involvement of organic bridging molecules in the mesoporous non-siliceous organic–inorganic hybrid materials makes the apparent strategies and the resultant mesoporosity much more abundant. Typically, template-free self-assembly and surfactant-mediated strategies stand for the two major ways to construct mesoporous non-siliceous hybrid materials.

Keywords Mesoporosity · Synthesis strategy · Template-free self-assembly · Surfactant-mediated · Pore periodicity · Crystallization

© The Author(s) 2015
Y.-P. Zhu and Z.-Y. Yuan, *Mesoporous Organic-Inorganic Non-Siliceous Hybrid Materials*, SpringerBriefs in Molecular Science,
DOI 10.1007/978-3-662-45634-7_3

3.1 Template-Free Self-assembly Synthesis Strategy

In recent years, researchers have paid much attention toward the synthesis of nanostructured porous hybrid materials through template-free self-assembly strategies, which do not require the use of preformed templates or structure-directing agents. These routes usually initiate the assembly from the interactions between the precursor molecules, and the ordered attachment allows the formation of porous morphologies.

Microporous organic–inorganic hybrid materials are often known as crystalline MOFs, which involve the strong and regular coordination of metal ions and organic linkage moieties, thus leading to porous framework structures. Figure 3.1 presents some typical organic acid ligands in obtaining mesoporous non-siliceous hybrid materials. Compared with typical microporous metal carboxylates, metal phosphonates have exhibited higher chemical and thermal stability due to the strong affinity and chelation of organophosphonic linkers to metal ions. Metal phosphonate hybrids often come up in the form of dense layered motifs, which have evolved into the field of organic–inorganic hybrids by appending organic pillars of the rigid inorganic layers [1]. It should be recognized that the pillars are too crowded and insufficient free space remains in the interlayer region, and no or poor porosity is expected to be present [2]. Several tactics have been adopted to create porosity in the metal phosphonate frameworks. The first route is the substitution of aryl biphosphonic acid by some non-pillaring groups, such as phosphoric, phosphors, and methylphosphonic acids, leading to the presence of interlayer pores and an increase in the surface area [3]. Although porous phosphonates can be obtained, the problem of this approach is that the replacement is random and uncontrollable, and the accurate structural characterization and a narrow pore size distribution are still challenges. Secondly, the geometry of a large and multidimensional polyphosphonic bridging molecules would disfavor the formation of the layered motif and thereby necessitate an open framework. A third approach would be to attach a second functional group to the phosphonate ligand to coordinate with the metal centers and disrupt the structure away from the layers [4].

Metal–sulfonate networks have been studied considerably less than other kinds of hybrid materials because of the relatively weak coordination interactions between the sulfonate anions and metal cations, making the frameworks insufficiently robust to sustain permanent porosity [5]. Metal sulfonates have been considered as potential analogues of layered metal phosphonates [6]. The rigid inorganic layers provide scaffolds of regular anchor points for pendant organic groups. In keeping with the theme of using larger cores with regard to the porous phosphonates to disperse the crossing sulfonate groups, the pillaring group, 1,3,5-tris(sulfomethyl)benzene could be envisioned to open channels between the layers of a metal sulfonate [7].

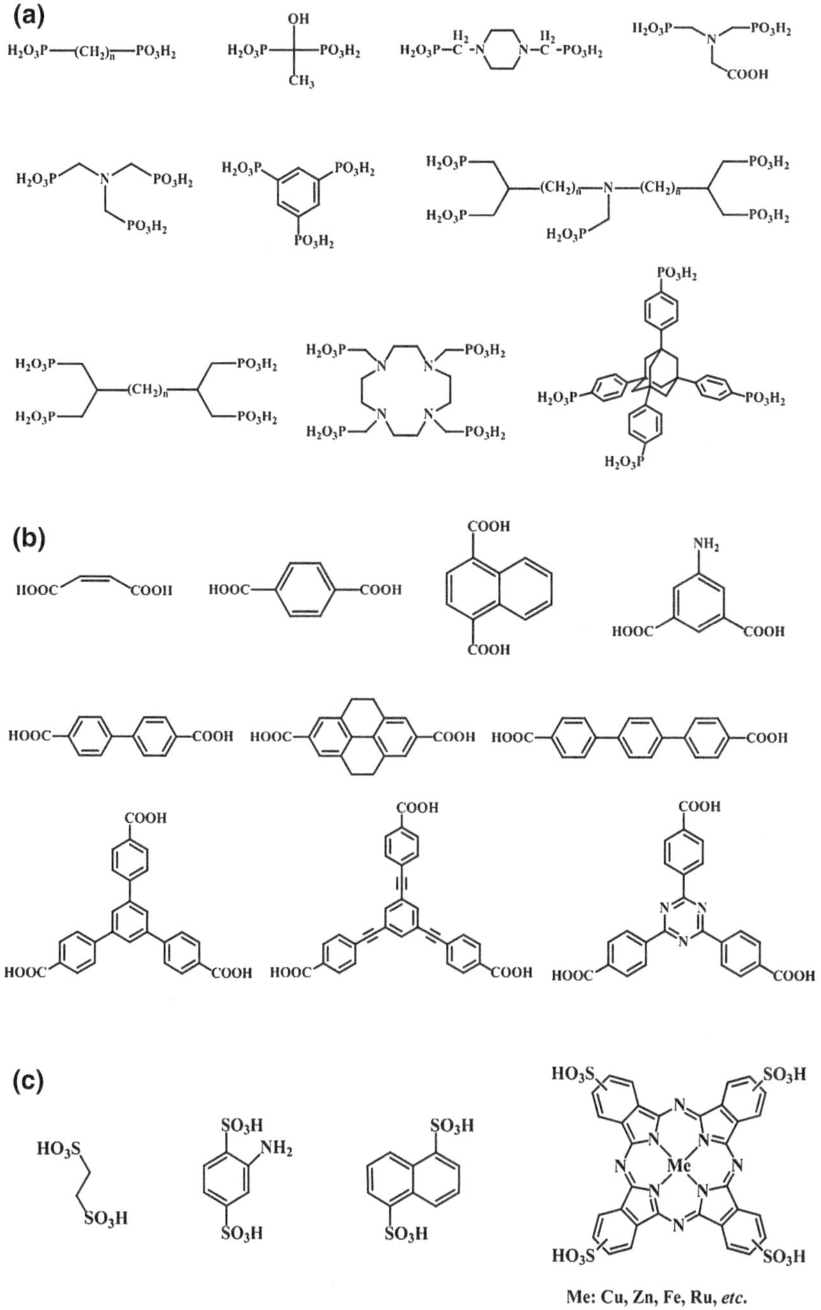

Fig. 3.1 Representative organic linkages for construction of mesoporous non-siliceous materials

3.1.1 Ligand Extension

Crystalline hybrids are constructed from the regular linkages of metal centers and organic groups, which can achieve well-defined pores, high surface area, and large porosity. Nevertheless, their pore sizes are typically restricted to the microporous range. Thus, the synthesis of mesoporous hybrids is envisioned to improve the transmission capability of the pores while facilitating practical applications that require the diffusion of bulky molecules [8]. Using extended ligands or bulky secondary building blocks is an apparent strategy. Disappointedly, linker expansion tends to result in the reduced surface areas and pore sizes due to the consequent interpenetrated structures and dramatically reduces the stability of the framework upon solvent removal from the porous hosts [9]. An elaborately designed ligand with hierarchical functional groups, 4,4′,4″-s-triazine-1,3,5-triyltri-p aminobenzoate, could be devised to extend the linkers while inhibiting interpenetration and reinforcing the framework against disintegration upon guest removal [10]. The mesoporous MOF was prepared through a one-pot solvothermal method, followed by stabilization at pH values around 3.0. The amino groups in the ligand were prearranged so that they would not participate in the framework formation but could accept protons after the network was generated, giving rise to a stable mesoporous MOF up to 300 °C. The N_2 sorption isotherm exhibited a typical type IV behavior, and the X-ray diffraction (XRD) data confirmed that the open channels were identical in size and as large as 22.5×26.1 Å. Similarly, Schröder et al. reported the synthesis of (3,24)-connected mesoporous framework NOTT-119 by adopting a nanosized $C3$-symmetric hexacarboxylate linker, presenting a high surface area of 4,118 $m^2 \, g^{-1}$ and pore sizes in the range of 2.4–4.5 nm [11]. The hybrid framework is stable up to 315 °C. When the dimension of the linker was enlarged beyond this point, the network could no longer hold stability to thermal treatment and suffered disruption of the structure owing to surface tension effects. A homologous series of palindromic oligophenylene derivatives terminated with α-hydroxy-carboxylic acid functions were targeted to afford linear and robust building blocks for expanding the pore size to up to 9.8 nm [12], which is the largest channel achieved via the ligand-expanding method to date. All members had non-interpenetrating structures and exhibited robust architectures, as evidenced by their permanent porosity and high thermal stability. This strategy for making MOFs with large pore apertures and avoiding the problem of interpenetration is to start with a framework in which one can maintain a short axis with the long organic links inclined to that axis. The short axis effectively eliminates the possibility of interpenetration because it is the distance between the links along that axis joining the secondary building units.

A multidimensional ligand, tetrakis-1,3,5,7-(4-phosphonatophenyl)adamantine (Fig. 3.2), which could impede the formation of a close-packed arrangement of organic molecules in an organic–inorganic hybrid framework, was used to fabricate mesoporous metal phosphonate materials [13]. The preparation was accomplished through a non-hydrolytic condensation process between the tetraphosphonic acid and titanium(IV) isopropoxide in DMSO, permitting the generation of mesopores. N_2 sorption experiments revealed the presence of

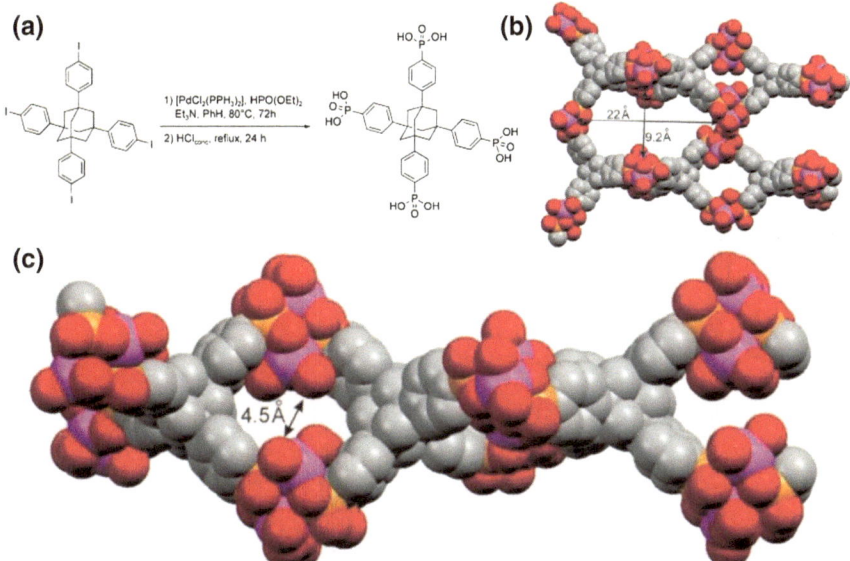

Fig. 3.2 a Synthesis of tetrakis-1,3,5,7-(4-phosphonatophenyl)adamantine. **b, c** Computer-stimulated model of titanium tetraphosphonate material. Reprinted with permission from Ref. [13]. Copyright 2006, Wiley-VCH

mesopores of about 3.8 nm accompanied with a surface area of approximately 550 m^2 g^{-1}. Additional XRD data suggested a paracrystalline material; an optimization of possible molecular arrangements was simulated that was in agreement with the experimental data. Furthermore, mesoporous vanadium phosphonates could also be obtained using the same method, exhibiting a BET surface area of 118 m^2 g^{-1} and a Barrett–Joyner–Halenda (BJH) pore diameter of 3.8–3.9 nm [14]. FT-IR and XPS as well as elemental analysis suggested that the phosphonate claw molecules were most likely connected in the form of ArP(O)O$_2$V$_2$O$_2$(O)PAr, which was effective for the aerobic oxidation of benzylic alcohol substrates.

[Mn$_2$(pdtd)$_2$(H$_2$O)$_4$]$_n$ · 5$_n$H$_2$O (pdtd = 3-(2-pyridyl)-5,6-diphenyl-1,2,4-triazine-4,4′-disulfonic acid) could be constructed by a solvothermal method preformed at 140 °C with methanol as the solvent [15]. Crystallographic analysis revealed acentric structure in which the Mn(II) centers are linked via sulfonate groups and chelating nitrogen atoms within the pdtd ligands to give a rare non-interpenetrating (10,3)-d framework with permanent helical cavities. As shown in Fig. 3.3, five pdtd ligands are linked by MnII ions to form a ten-membered nanoscale loop (Mn1–Mn2: 9.1 Å; Mn1–Mn1C: 25.5 Å). The loops are further connected via edge-sharing to form a 3D framework with large channels that run parallel to the *a* axis. Disappointedly, studies concerning mesoporous metal sulfonates through the ligand extending way have been scarce reported to the best of our knowledge. The unavoidable constrainment is that the weak coordination of sulfonate functional groups makes them difficult to maintain the permanent porosity and periodic crystallite structures.

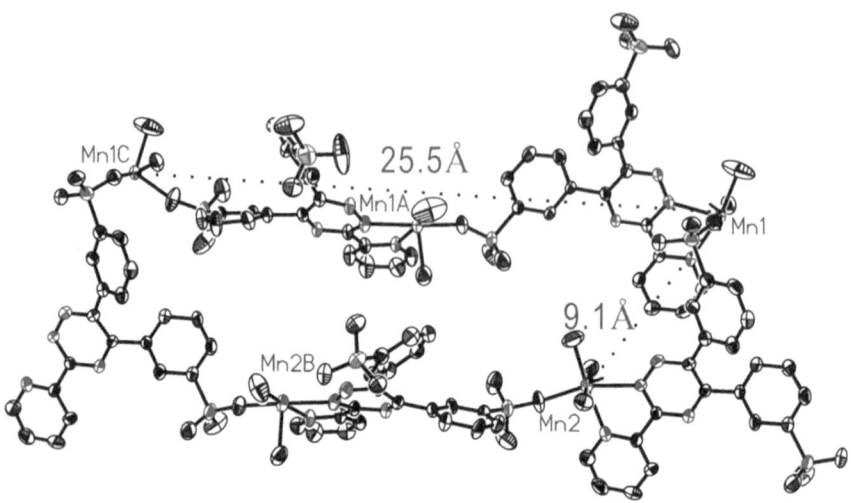

Fig. 3.3 View of the pentanuclear loop in 1. Symmetry codes: A: x, $-y$, $z + 1/2$; B: x, $-y - 1$, $z + 1/2$; C: x, y, $z + 1$. Reprinted with permission from Ref. [15]. Copyright 2011, Wiley-VCH

3.1.2 Microemulsion Method

Ligand extension strategy is an effective and simple way to expand micropores to the mesoporous regime in hybrid materials; though the most complicated organic linkers are not commercially available and in order to gain them, complicated fabrication processes are generally required. More recently, a new method to prepare mesoporous hybrid metal phosphonates, via employing microemulsions, was reported [16]. Mesostructured pores of several nanometers in size existed in the vicinity of the surface in a wormhole-like assembly (2.5–5.8 nm), whereas in the core, close to the wormhole-like mesoporous surface layers of the particle, a new mesocellular foam structure (8–10 nm) similar to the previously reported mesostructured cellular foam (MCF) silica materials was observed (Fig. 3.4). During the period of synthesis, hydrolysis of titanium tetrabutoxide in EDTMP aqueous solution resulted in the rapid formation of nanometer-sized titanium phosphonate and butanol molecules at the same time. Thereafter, microemulsion drops formed in the multicomponent system of alkoxide/organophosphonate–alcohol–water while stirring. The phosphonate sols aggregated along with the microemulsions, evolving to a mesocellular foam structure. The interactions between the sols caused the formation of mesostructured nanoclusters of several nanometers in size. At this stage, due to the presence of a large amount of butanol by-products, the reaction mixture was transferred to phosphonate-based mesophases and water–alcohol domains by microphase separation, induced by aging, leading to discrimination of them and even some macrovoids [17]. If 1-hydroxy-ethylidene-1,1-diphosphonate (HEDP) was chosen as the organic precursors, according to microemulsion methodology, the interfacial polymerization

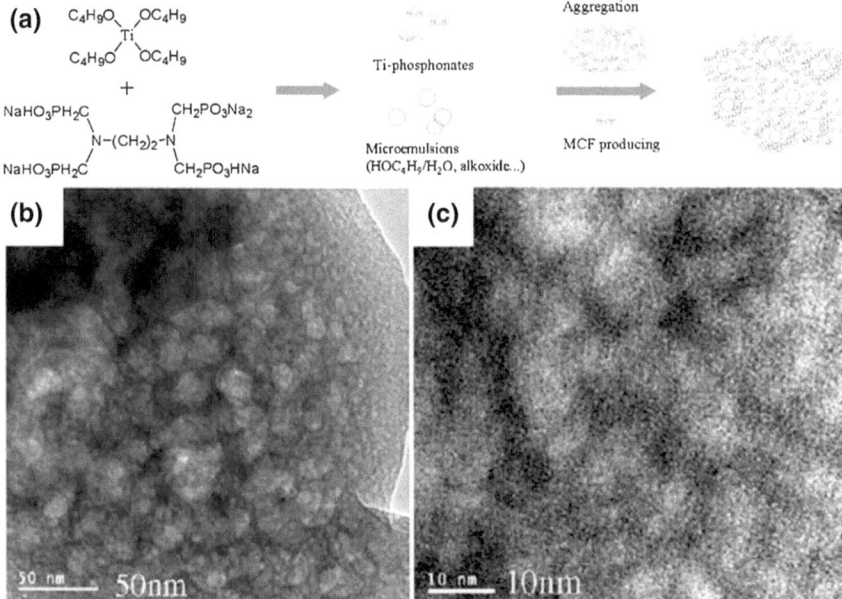

Fig. 3.4 **a** Proposed formation mechanism of titanium phosphonates with both wormhole-like mesopores and MCF structures, and **b, c** the corresponding TEM images. Reprinted with permission from Ref. [16]. Copyright 2009, Elsevier

of titanium phosphonate sols and titanium–oxo clusters resulted in the formation of mesostructured hybrid nanorods with a length of 80–150 nm and a thickness of 18–38 nm [18], which further aggregated along with the microemulsions to give a hierarchical macroporous structure. In this process, phase separation might take place in the growing aggregates of titanium phosphonate-based mesophases and water/alcohol domains, leading to the creation of huge, sporadic macrochannels in hierarchical macroporous networks. The multipoint BET surface area was 257 m^2 g^{-1}, accompanied with a BJH pore size of 2.0 nm and a total pore volume of 0.263 cm^3 g^{-1}. The fabrication of other metal phosphonates with different organic bridging groups and hierarchical nanostructures can also be expected, wherein metal alkoxide was preferred owing to the special solgel reaction process.

3.1.3 Nanocrystal Self-assembly

The preparation of single-sized porous hybrid materials from angstroms to micrometers has been an interesting research area in the past several decades [19]. However, materials presenting multiscale or multimodal porosity receive even higher interest due to the enhanced mass transport through the larger pores and maintenance of a specific surface area on the level of fine pore systems, which is a central point of many processes concerning accessibility, especially catalysis, adsorption, optics, and sensing [20, 21]. Expanding the pore width of the crystalline organic–inorganic hybrid via a ligand extension method has been demonstrated [22]. However, the creation of larger pores through ligand extension usually involves the expense of the loss of micropores. Moreover, the ordered nanostructures or pores could only be retained for small mesopore sizes or constricted cages [23]. Thus, a surfactant-free methodology was performed for the synthesis of Zn-MOF-74 with hierarchical micro-/mesoporosity using 2,5-dihydroxy-1,4-benzenedicarboxylate [24]. The synthesis proceeded at room temperature to restrict the crystallization, and finally nanosized MOF-74 crystals were formed. The acetate from inorganic precursors could accelerate the crystal generation through a ligand exchange process [25] that would not be expected to happen with the conjugate bases of strong acids (e.g., Cl^-, SO_4^{2-}, and NO_3^-). The precipitate materials were composed of discrete MOF nanoparticles embedded in an amorphous matrix, which exhibited large interparticle mesopores in the 2–20 nm range and intraparticle micropores with a maximum at about 1.1 nm confirmed by N_2 sorption and transmission electron microscopy (TEM) observation.

Aerogels are well known because of their versatile porosity, low density, and high internal surface area, but the design of aerogels is still rudimentary due to a large disorder in the structure and the insufficient prediction of gelation behavior [26]. Noticeably, metal–ligand coordination and other supramolecular forces (e.g., H-bonding, π–π stacking, and van der Waals interactions) are emerging as effective driving forces in gelation, offering metal–organic gels (MOGs) as a novel class of functional soft materials [27, 28]. On the basis of the inherent correlations between MOFs and MOGs, Li et al. reported a general synthetic route for the fabrication of hierarchically porous metal–organic aerogels (MOAs) via MOG formation from the self-assembly of a precursor (Fig. 3.5) [29]. Typically, the strong metal–ligand coordination interactions impel the metal ions and organic linkages to assemble into MOF clusters, which then polymerize to MOF nanoparticles with well-defined microporosity. Under the reaction conditions that favored the consistent epitaxial growth or oriented attachment [30, 31], the further crystallization of MOF subunits could lead to bulky MOFs. Moderate heating represented a key factor in mismatched growth or cross-linking of preformed MOF particles and then triggering the gelation of proper solvents (mainly ethanol). The careful removal of the solvents by sub-/supercritical CO_2 extraction left hierarchically porous MOAs based on MOGs. The N_2 sorption isotherms are between type I, characteristic of microporous materials, and type IV, indicative of mesoporous materials. The texture and porosity could be easily adjusted by changing precursor concentrations.

Zhao and coworkers reported the synthesis of stable bicontinuous hierarchically porous MOFs (ZIF-8 and HKUST-1) with the assistance of a coordination regulating agent [32]. Two functional block co-oligomer templates were used independently, i.e., poly(styrene)-block-poly(4-vinylpyridine) and poly(styrene)-block-poly(acrylic acid). Two prototypical MOFs, ZIF-8 and HKUST-1, were selected to demonstrate our approach. The resulting materials resembled the microstructures of bicontinuous silica aerogels, composed of a branched fibrous

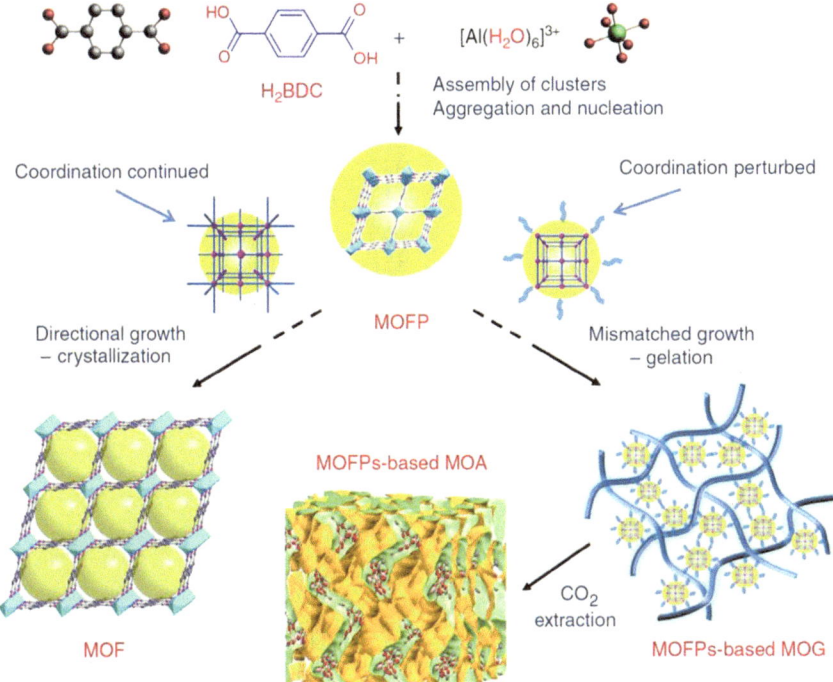

Fig. 3.5 Schematic representation of the formation of highly crystalline MOF versus MOA. Reprinted with permission from Ref. [29]. Copyright 2013, Nature Publishing Group

network of interconnected about 40-nm-sized microporous nanocrystallites. Besides the intrinsic micropores, the prepared ZIF and HKUST-1 exhibited aerogel-like textural interparticular voids from 2 to 100 nm, encompassing the mesoporous and macroporous regions that were absent in the single crystal forms. The crown ether molecules complex the Zn^{2+} and Cu^{2+} at first and then gradually release the metal ions upon reaction with the organic ligands. The reduced nucleation and crystallization rates lead to improved crystallinity of the resultant MOFs. Significantly, the pH of the template solutions influences the self-assembly of block co-oligomers, the capture of metal ions, and the protonation/deprotonation of the ligands.

A facile phosphate-mediated self-assembly methodology has been carried out to prepare mesoporous nickel phosphate/phosphonate hybrid microspheres [33]. The hybrid microspheres are composed of not only crystalline nickel phosphate but also amorphous nickel phosphonate (Fig. 3.6), and the formation of crystalline $Ni_3(PO_4)_2$ nanoparticles played crucial roles in the formation of these hybrid microspheres. When phosphoric acid and nickel salts were mixed under vigorous stirring, uncountable $Ni_3(PO_4)_2$ nanoparticles of a few nanometers were emerged instantaneously. Nickel phosphonate clusters were formed after the addition of ethylene diamine tetra(methylenephosphonic acid) (EDTMP) to react with the

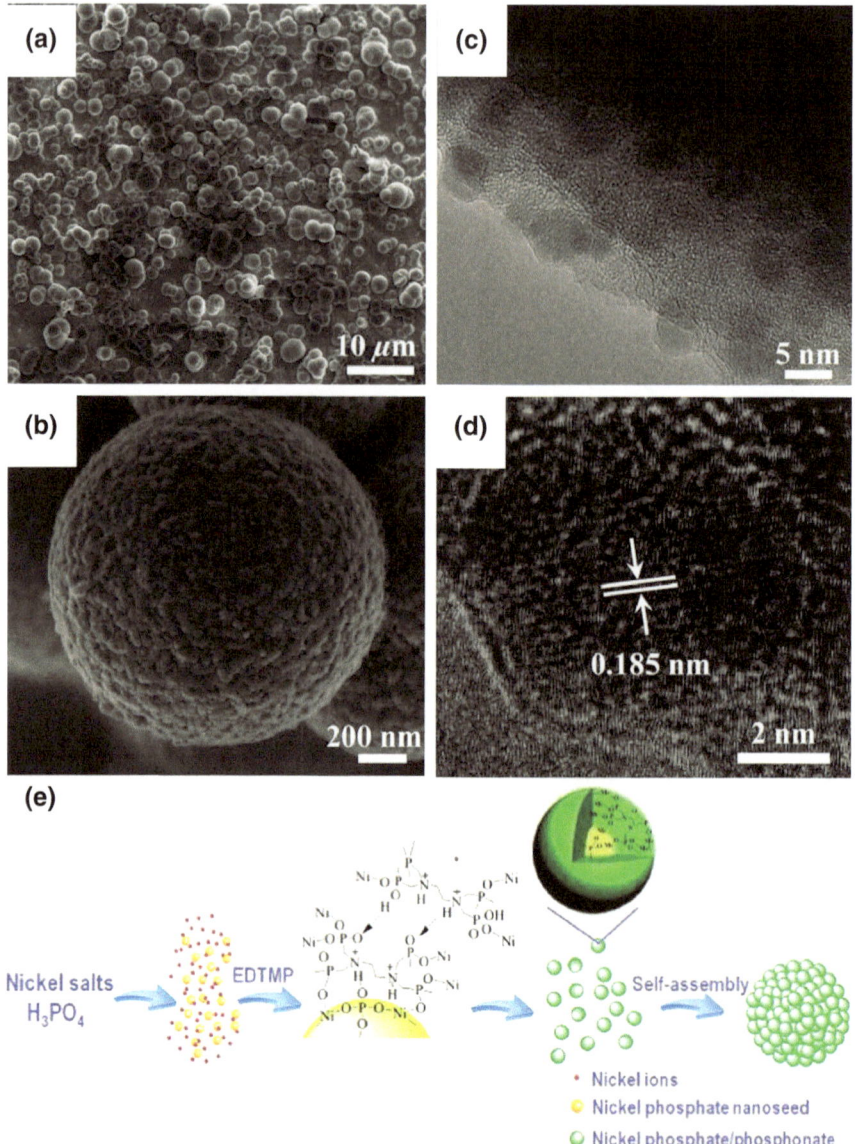

Fig. 3.6 **a, b** SEM and **c, d** TEM images of the NiPPH material, crystalline interplanar spacing of 0.185 nm is attributable to the (510) plane of $Ni_3(PO_4)_2$. **e** Formation and preparation process of the nickel phosphate/phosphonate hybrid microspheres. Reprinted with permission from Ref. [33]. Copyright 2014, Royal Society of Chemisty

rest of Ni^{2+}. These phosphonate clusters attached and wrapped the phosphate "nanoseeds" to generate core-shell-like secondary building blocks (SBB) under the driving of intermolecular interaction, as depicted in Fig. 3.6. Thereafter, phosphate/phosphonate SBB would self-assemble to hierarchical hybrid microspheres to minimize the interfacial energy. Distinct from the high solubility of nickel chloride, nitrate, and sulfate in water, nickel phosphate is insoluble. If phosphoric acid was substituted by hydrochloric, nitric, and sulfuric acids, the resultant products were amorphous nickel phosphonate nanoparticles. The hybrid microspheres could also be obtained when Na_2HPO_4, NaH_2PO_4, and Na_3PO_4 were employed as inorganic phosphors sources, suggesting that $Ni_3(PO_4)_2$ involved in the self-assembly procedure. The selection of the right phosphonic precursors is considerably significant to determine the resultant micromorphology. In this work, the tetraphosphonic claw molecules containing pyridinic nitrogen could protonate in acid environment and thus form zwitterions, which could contribute to the hydrogen-bonding interactions between the phosphate nanoparticles and the slowly formed phosphonates. A kind of biphosphonic acid (HEDP) without pyridinic nitrogen component was tried to substitute EDTMP to prepare hybrid microspheres, but failed, signifying the positive roles of proper weak interaction. Another N-containing phosphonic acid (bis(hexamethylene triamine penta(methylene phosphonic acid)), BHMTPMP) with similar molecular structure to EDTMP but longer alkyl chains ($-[CH_2]_6-$) was used as well, and spherical nanoparticles were gained, which might be due to that the strong hydrophobic interaction between the hexamethylenetriamine bridges could disturb the intermolecular interactions and the self-assembly process. The resultant nickel phosphate/phosphonate material presents great potential in CO_2 capture and heavy metal ion removal.

Mesoporous even hierarchically porous non-siliceous hybrids could be obtained through a self-assembly approach in the absence of any templates or surfactants. No matter whether using linkage extension, microemulsion, or nanocrystal self-assembly methodologies, they usually involve weak interactions among the hybrid nanobuilding blocks. Other interactions such as hydrogen bonds, hydrophobicity–hydrophobicity interactions, and π–π stacking can also direct the spontaneous formation of mesoporous non-siliceous materials with fascinating porosity, structures, and stability, which are mainly dependent on the synthesis conditions. This means that the synthesis processes are difficult to control to some extent. Solgel chemistry presents an alternative route to synthesize mesoporous even hierarchical porous hybrid materials, wherein the solgel processes are directly associated with the precursor species and solvents.

3.2 Surfactant-Mediated Synthesis Strategy

Although surfactant- or template-free approaches have been proven to be valuable methods to obtain hybrid organic–inorganic materials presenting porosity from micropores to macropores, these methods cannot afford the valid capability to adjust the porosity, texture, and even morphologies on demand. As to some kinds of potential applications, such as separation and recognition of large molecules, catalysts and catalyst supports, and dye adsorption, the existence of mesopores is much more preferable than micropores and macropores to some certain extent. The use of supramolecular templates to synthesize ordered mesoporous materials

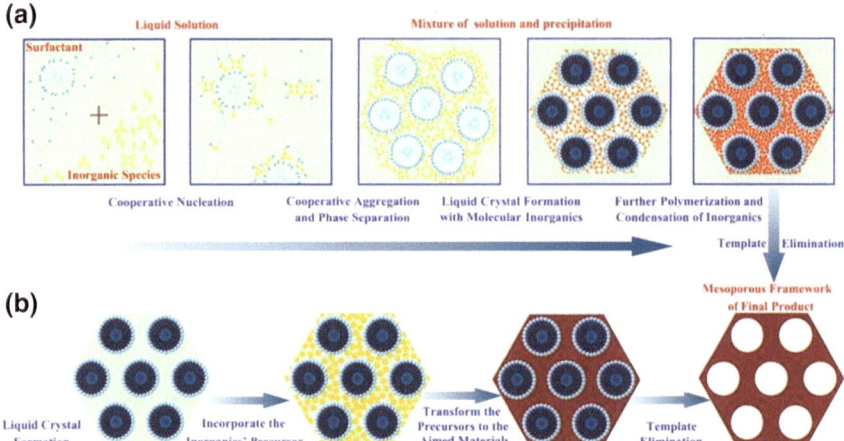

Fig. 3.7 Schematic illustration of **a** hard- and **b** soft-templating methods to synthesized mesoporous materials

has received increasing attention in the last few decades, due to the periodically aligned pore systems, uniform pore size in the mesoscale range, high surface area, controllable mesophase, and abundant framework compositions, and has found diverse applications in the fields of adsorption, separation, catalysis, biosensing, and energy storage and conversion [34, 35]. The surfactant-induced route for the formation of mesopores can be classified into two types (Fig. 3.7). The first ones are termed as hard templates or nanocasting, which are the prepared mesoporous materials with solid frameworks, including carbons, polymer beads, and silicas. However, the prepared materials often have a wider pore size distribution than that of the pristine replicas, and multiple preparation procedures with costly hard templates make it expensive, complicated, and consequently unsuitable for large-scale production and industrial applications. Meanwhile, the template removal always involves the use of strong acids, bases, and high-temperature calcination, which cause it to be favorable for the synthesis of special mesoporous materials, for instance, metal sulfides and oxides [36, 37], carbons [38, 39], and silicon carbides [40], while not suitable for the cases of hybrid materials. The soft-templating methodology, which is usually referred to as "soft" molecules including cationic surfactants $C_nH_{2n+1}N(CH_3)_3Br$ ($n = 8$–22) and nonionic surfactants of amphiphilic poly(alkylene oxide) triblock copolymers [e.g., PEO-PPO-PEO (PEO = poly(ethylene oxide), PPO = poly(propylene oxide))] and oligomeric alkylethylene oxides, has received much attention. In comparison with the nanocasting method, the entire procedure of soft-templating is low costing, facile, convenient, effective, and promising for large-scale production. More importantly, the mesophase formation depends on the temperature, type of solvent, ionic strength and pH, and the nature of the template molecules (hydrophobic/hydrophilic volume ratio, hydrophobic length, etc.), which make the pore structure and surface properties easily tuned. As to hybrid frameworks containing organic components, the soft-templating approach is much more appropriate due to the modest preparation conditions to protect the hybrid frameworks, relative simplicity and environmental friendliness. Thus, the preparation of mesoporous hybrid materials through the soft-template strategy will be discussed in detail in the following sections.

3.2.1 Synthesis Mechanism

Since the liquid–crystal templating theory was proposed, there have been an unprecedented number of studies concentrated on the synthesis, modification, and application of mesoporous materials. Mesoporous inorganic–organic hybrids are no exception. The selection of surfactants is a key factor. Proverbially, surfactants consisting of hydrophilic heads and hydrophobic tails can assemble into micelles at a concentration higher than the critical micelle concentration under the driving force of hydrophobic interactions. A lyotropic liquid can provide an organized scaffold (Fig. 3.8). The formed oligomers from the condensation and polymerization of organic–inorganic precursors grow around the arranged surfactant micelles driven by the interactions (e.g., electrostatic forces and hydrogen) between the surfactant molecules and the oligomers. After a further condensation and polymerization, the surfactants can be removed, leaving a mesoporous structure.

Frequently and commercially used surfactants can be classified into cationic, anionic, and nonionic surfactants. Quaternary cationic surfactants (e.g., cetyltrimethylammonium bromide (C_{16}TABr) have excellent solubility and high critical micelle temperature values and are generally efficient for the synthesis of ordered mesoporous materials. Nonionic surfactants are available in a wide variety of different chemical structures. They are widely used in industry because of their attractive characteristics, such as low cost, non-toxicity, and biodegradability. Nonionic surfactants have rich phase behaviors and low critical micelle temperature values and have become more and more popular and powerful in the synthesis

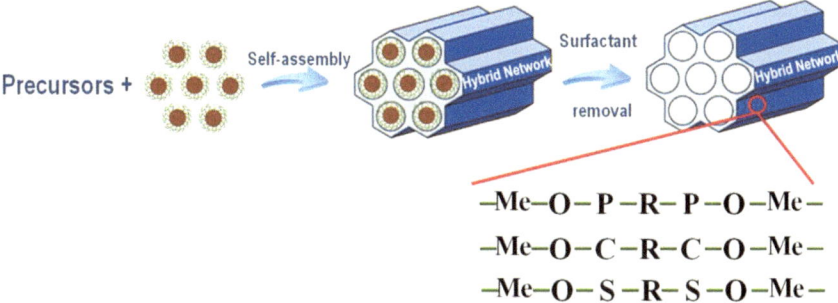

Fig. 3.8 General synthesis mechanism for the formation of mesoporous non-siliceous hybrids through soft-templating strategy

Table 3.1 Summary of surfactant-mediated synthesis routes and the corresponding surfactants

Route	Interaction	Typical pH condition
S^+I^-	Electrostatic and coulomb force	Basic media
S^-I^+	Electrostatic and coulomb force	Acidic media
S^0I^0	Hydrogen bond	Neutral condition
$S^+X^-I^+$	Electrostatic and coulomb force, double-layer hydrogen bond	Acidic media
$S^-X^+I^-$	Electrostatic and coulomb force, double-layer hydrogen bond	Basic media
$(S^0H^+)X^-I^+$	Electrostatic and coulomb force, double-layer hydrogen bond	Acidic condition

of mesoporous solids. Commercially available Pluronic F127 ($EO_{106}PO_{70}EO_{106}$) and P123 ($EO_{20}PO_{70}EO_{20}$), and Brij 56 ($C_{16}EO_{10}$) are often used. However, as to anionic salt surfactants including carboxylates, sulfates, sulfonates, and phosphates, there are few reports of their use in the synthesis of inorganic–organic hybrid non-silica-based materials, though a family of mesoporous silica structures (AMS-*n*) has been prepared under basic conditions by employing anionic surfactants as structure-directing agents with the assistance of aminosilanes [41], based on the charge-matching effect.

According to the synthetic conditions, the chemical diversity of the composite mesoporous materials has been expanded during the last few decades, with many synthesis pathways being demonstrated during the nucleation of the composite phase, such as direct surfactant–inorganic interaction (S^+I^-, S^-I^+, S^0I^0) and mediated interaction ($S^+X^-I^+$, $S^-X^+I^-$, $(S^0H^+)X^-I^+$) (S^+ = surfactant cations, S^- = surfactant anions, I^+ = inorganic precursor cations, I^- = inorganic precursor anions, X^+ = cationic counterions, and X^- = anionic counterions) [42]. To yield mesoporous materials, it is important to adjust the chemistry of the surfactant head groups that can fit the requirement of the components. Table 3.1 lists the main synthesis routes and the corresponding surfactants. Furthermore, the mesophase behavior and pore width are dominated by the liquid–crystal scaffolds. Some typical examples of ordered mesoporous metal phosphonates are summarized in Table 3.2, which contains the experimental parameters, textual properties, and mesophases.

3.2.2 How to Effectively Obtain Periodic Mesoporosity

Hybrid oligomers can be either generated during the reaction process or preformed before being assembled with the surfactant micelles. The key factors for the assemblies to form a periodic mesophase include the control of the aggregation of precursors, the presence of sufficient interactions between oligomers/precursors and surfactants, and in turn, a proper size and charge of suitable building blocks [43].

Table 3.2 Summary of synthetic strategies and physicochemical properties of representative mesoporous metal phosphonates

Metallic precursor	Phosphonic acid[a]	Strategy	Surface area $(m^2\,g^{-1})$	Pore volume $(cm^3\,g^{-1})$	Pore size (nm)	Surfactant	Additives[b]	Mesophase	Ref.
Ti(O)(i-PrO)$_4$	TPPhA	Ligand extension	557	0.42	3.8	–	–	–	[13]
V(O)(i-PrO)$_3$	TPPhA	Ligand extension	118	–	3.8–3.9	–	–	–	[14]
Ti(O)(n-Bu)$_4$	EDTMP	Microemulsion	86	0.074	2.5–5.8, 8–10	–	–	–	[16]
Ti(O)(n-Bu)$_4$	HEDP	Microemulsion	257	0.263	2.0	–	–	–	[18]
NiCl$_2$	EDTMP	Nanocrystal self-assembly	267	0.191	5.3	–	–	–	[33]
Al(O)(i-PrO)$_3$	PDP, $n = 1$	–	738	0.32	1.8	C$_{16}$TACl	–	Hexagonal	[46]
AlCl$_3$	PDP, $n = 1$	EISA	788	0.44	2.2	C$_{16}$TACl	–	Hexagonal	[47]
AlCl$_3$	PDP, $n = 2$	EISA	708	0.32	1.9	C$_{16}$TACl	–	Hexagonal	[47]
AlCl$_3$	PDP, $n = 3$	EISA	217	0.27	3.0	Brij 58	–	Hexagonal	[48]
AlCl$_3$	PDP, $n = 3$	EISA	172	0.59	7.3	F68	–	Hexagonal	[48]
AlCl$_3$	PDP, $n = 2$	EISA	309	0.71	9.4	P123	–	Hexagonal	[48]
AlCl$_3$	PDP, $n = 2$	EISA	337	0.79	11.6	F127	–	Hexagonal	[48]
Al(O)(n-Bu)$_3$	PDP, $n = 2$	Artrane route	675	0.63	3.3	C$_{16}$TABr	–	Hexagonal	[51]
TiCl$_4$	EDTMP	Hydrothermal and EISA	1,066	0.83	2.8	Brij 56	–	Hexagonal	[52]
TiCl$_4$	HEDP	Hydrothermal and EISA	1,052	0.58	2.6	C$_{16}$TABr	–	Cubic	[65]
Al(O)(i-PrO)$_3$	PDP, $n = 1, 2$	EISA	–	–	–	C$_{16}$TACl	–	Lamellar	[66]
AlCl$_3$	EDTMP	Microwave-assisted hydrothermal	498	0.61	1.7, 7.5	F127	–	Hexagonal	[73]
TiCl$_4$	EDTMP	Microwave-assisted hydrothermal	522	0.63	1.4, 7.2	F127	–	Hexagonal	[73]

(continued)

Table 3.2 (continued)

Metallic precursor	Phosphonic acid[a]	Strategy	Surface area ($m^2 g^{-1}$)	Pore volume ($cm^3 g^{-1}$)	Pore size (nm)	Surfactant	Additives[b]	Mesophase	Ref.
$ZrCl_4$	EDTMP	Microwave-assisted hydrothermal	513	0.64	1.5, 7.1	F127	–	Hexagonal	[73]
VCl_4	EDTMP	Microwave-assisted hydrothermal	538	0.68	1.3, 7.1	F127	–	Hexagonal	[73]
$AlCl_3$	PDP, $n = 1$–4	Spray-drying	62–170	0.12–0.42	30–200	PS-b-PEO	–	–	[75]
$AlCl_3$	PDP, $n = 2$	Spray-drying	251	0.66	10.5	F127	–	Hexagonal	[77]
$AlCl_3$	PDP, $n = 2$	Spray-drying	189–315	0.62–0.77	11.2–15.0	F127	TMB	Hexagonal	[77]
$AlCl_3$	PDP, $n = 2$	Spray-drying	257–307	0.77–1.32	12.8–21.8	F127	TIPBz	Hexagonal	[77]
$ZrOCl_2$	HEDP	Hydrothermal	702	0.86	3.6	$C_{16}TABr$	–	Worm-like mesopores	[79]

[a]*PDP* propylene diphosphonic acids; [b]*TMB* 1,3,5-trimethylbenzene; *TIPBz* 1,3,5-triisopropylbenzene

For the synthesis of ordered mesoporous siliceous materials, it is relatively easy to gain chargeable hydrated silicate oligomers due to the undertaken hydrolysis of the inorganic precursors in a certain pH environment. In contrast, the uncontrollable hydrolysis and condensation for most non-siliceous inorganic precursors and the tendency to form crystalline products make it difficult to apply the liquid–crystal templating mechanism to produce the stable and accessible long-range periodic mesostructures. The synthesis system is rather complicated; for instance, multi-inorganic precursors, solvents, acids or base to maintain a steady pH, a retainable chemical integrity, and preformed mesostructures are needed during the process of surfactant removal. In order to obtain ordered mesoporous hybrid materials, an appropriate system is necessary to prohibit or reduce the hydrolysis of inorganic sources and the coordination rates of metal ions and organic groups and to enhance the interactions between the surfactant scaffolds and the charged oligomers.

A family of alkyl pyrazinium surfactants that were tethered to Prussian blue (PB) precursors could enable the isolation of a kinetically controlled mesostructure [44]. Further reactions with $Na_3[Fe(CN)_5NH_3] \cdot 3H_2O$ gave an amphiphilic intermediate that templated the formation of mesostructured PB analogues. However, the chemical bonds between the surfactant ionic heads and PB-type inorganic skeleton made it difficult to remove the surfactants. Direct cooperative self-assembly of metal ions, cationic surfactants, and ligands with weak coordination acting sites to induce surfactants to overcome the lattice stability was accomplished by Li et al. [45], resulting in mesostructured MOFs with an amorphous wall. Also, the surfactant molecules could not be removed from the final MOF products.

A high acid reaction system has been proven as an efficient way to lower the hydrolysis of metal sources owing to the instant formation of large amounts of positively charged hybrid oligomers. Kimura prepared highly ordered mesoporous aluminum organophosphonates using alkyltrimethylammonium surfactants [46, 47]. In the ethanol–water system, ordered hexagonal mesostructures could be gained through the reaction of aluminum chloride and alkylene diphosphonic acids under highly acidic conditions. However, the XRD patterns of the mesoporous materials showed a low ordering of the mesostructures. Elemental analysis indicated the presence of Cl^- anions in the frameworks. This revealed that the mesostructured materials were conducted through the $S^+X^-I^+$ or $(S^0H^+)X^-I^+$ pathway, demonstrating the impurity of the hybrid framework. The surfactant molecules accommodated in the mesopores could not be extracted by conventional acid treatment due to the less condensed and easily hydrolyzed networks, and thus, calcination at 400 °C was carried out. Oligomeric surfactants or triblock copolymers could also be used for the preparation of mesoporous aluminum phosphonates [48].

Addition of organic solvents or organic chelates is another alternative approach to inhibit the hydrolysis. Pure periodic mesoporous aluminum phosphonates and diphosphonates could be obtained by using aluminum "atrane" complexes, and methylphosphonic and ethylenediphosphonic acids through an S^+I^- surfactant-assisted cooperative mechanism by means of a one-pot preparative procedure [49, 50]. A soft chemical extraction procedure enabled the opening of the pore system of the parent mesostructured materials by exchanging the surfactant without the collapse of the mesostructure. The BET surface area of the mesoporous hybrid could reach up to 793 cm^2 g^{-1}, accompanied with a narrow pore width distribution around 2.7 nm. This procedure was on the basis of the use of $C_{16}TABr$ as a structure-directing agent and 2,2',2''-nitrilotriethanol as the complexing polyalcohol, which had proven its capability in controlling the rates of hydrolytic reactions of aluminum species in water–phosphoric acid media and the subsequent process of self-assembly in the presence of surfactant aggregates [51]. The hybrid nature of the pore wall can be modulated continuously from organic-free mesoporous aluminum phosphates (ALPOs) up to total incorporation of organophosphorus entities (mesoporous phosphonates and diphosphonates). The organic functional groups become basically attached to the pore surface or inserted into the ALPO framework (homogeneously distributed along the surface and inner pore walls) depending on the use of phosphonic or diphosphonic acids, respectively.

The successful preparation of periodic mesoporous titanium phosphonate (PMTP-1) with bridged organic linkers inside the framework was achieved by Ma et al. [52] via an autoclaving process followed by an evaporation-induced self-assembly (EISA) strategy (Fig. 3.9). To slow down the hydrolysis of titanium tetrachloride, the metallic precursors were dissolved in the ethanol previously to form $TiOCH_2CH_3$ complexes [53, 54]. A cryosel bath was used to create low-temperature conditions as well to reduce the condensation speeds of the reactants. This could avert the generation of large titania or titanium phosphonate aggregations during the reaction process. The highly ordered mesostructures were obtained when a moderately acidic pH value was sustained, according to the $(S^0H^+)X^-I^+$ mechanism. This was probably due to the newly formed gel being partially damaged in the strong acid system and that alkaline conditions led to a fast hydrolysis rate. The surface area, pore size, and pore volume were 1,066 m^2 g^{-1}, 2.8 nm, and 0.83 cm^3 g^{-1}, respectively. This mechanism could be extensively applied to the formation of a series of periodic mesoporous metal phosphate and phosphonate materials with different structural phases in the presence of nonionic surfactants in acidic media [55, 56].

Ionic liquids (ILs), considered as tunable and environmentally friendly solvents, have attracted a lot of interest in the synthesis of novel materials. Zhang et al. [57] synthesized well-ordered mesoporous MOF nanospheres constructed by a microporous framework in a system of ILs–surfactant combined with superficial CO_2 (Fig. 3.10). The IL and surfactant chosen were 1,1,3,3-tetramethylguanidinium acetate (TMGA) and N-ethyl perfluorooctylsulfonamide (EtFOSA), respectively. It has been shown that TMGA/EtFOSA/CO_2 microemulsions could be formed [58]. The surfactant molecules self-assembled into cylindrical micelles with the fluorocarbon chain directed toward the inside of the micelles, and CO_2 existed as a core of the micelles. Thereafter, the Zn(II) ions and 1,4-benzenedicarboxylic acid linked facilely around the formed micelles. Thus, the mesoporosity was generated from the templating effect of the surfactants, and the microporosity was related to the intracrystalline cavities. The calculated sizes of mesopores and micropores were 3.6 and 0.7 nm, respectively, as well as a total surface area

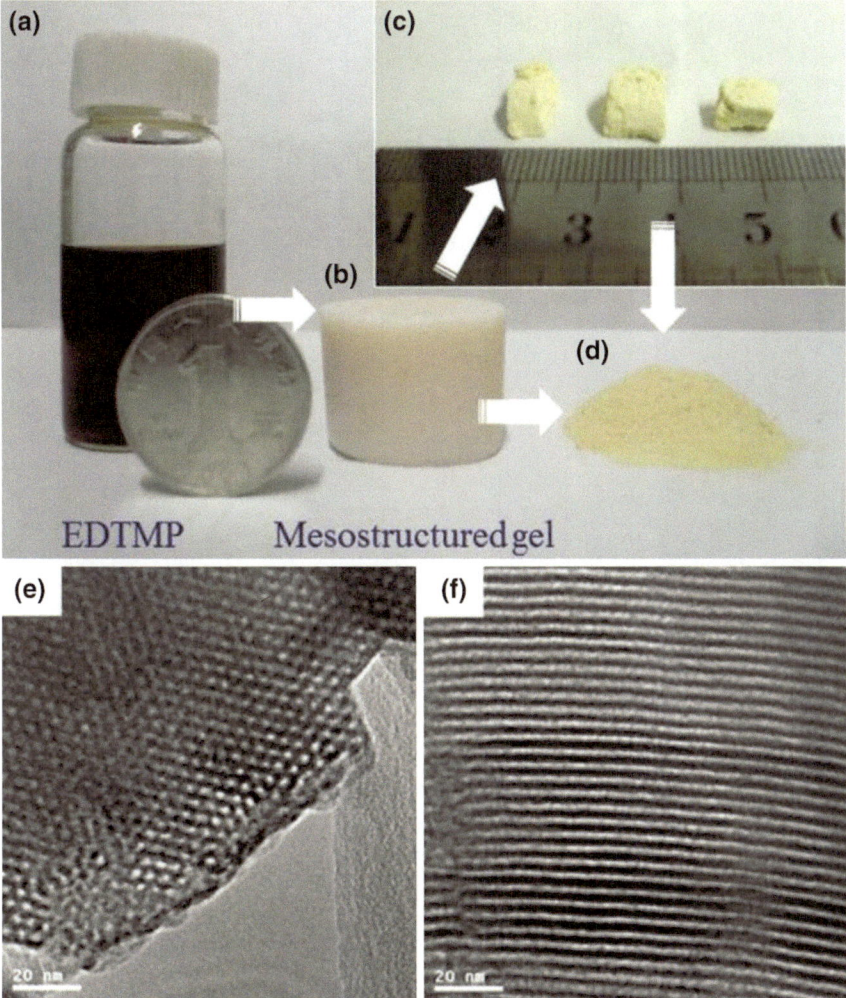

Fig. 3.9 Photographs of **a** EDTMP solution, **b** as-synthesized gel before surfactant extracted, **c** mesostructured monoliths after extraction, and **d** final mashed powder. TEM images of the hexagonal mesoporous titanium phosphonate hybrid material (**e**, **f**). Reprinted with permission from Ref. [52]. Copyright 2010, Royal Society of Chemistry

of 756 m^2 g^{-1}. The corresponding crystal structure could not be identified mainly owing to the small size of the nanoparticles.

An abundant variety of strategies have been used to effectively synthesize mesoporous non-siliceous hybrids. To efficiently mediate the coordination and condensation of organic linkers and inorganic species while increasing the interactions between surfactant scaffolds and the formed oligomers is the key factor. The removal of surfactant molecules to leave mesovoids in the framework was preferred for using moderate measures, including extraction by organic polar

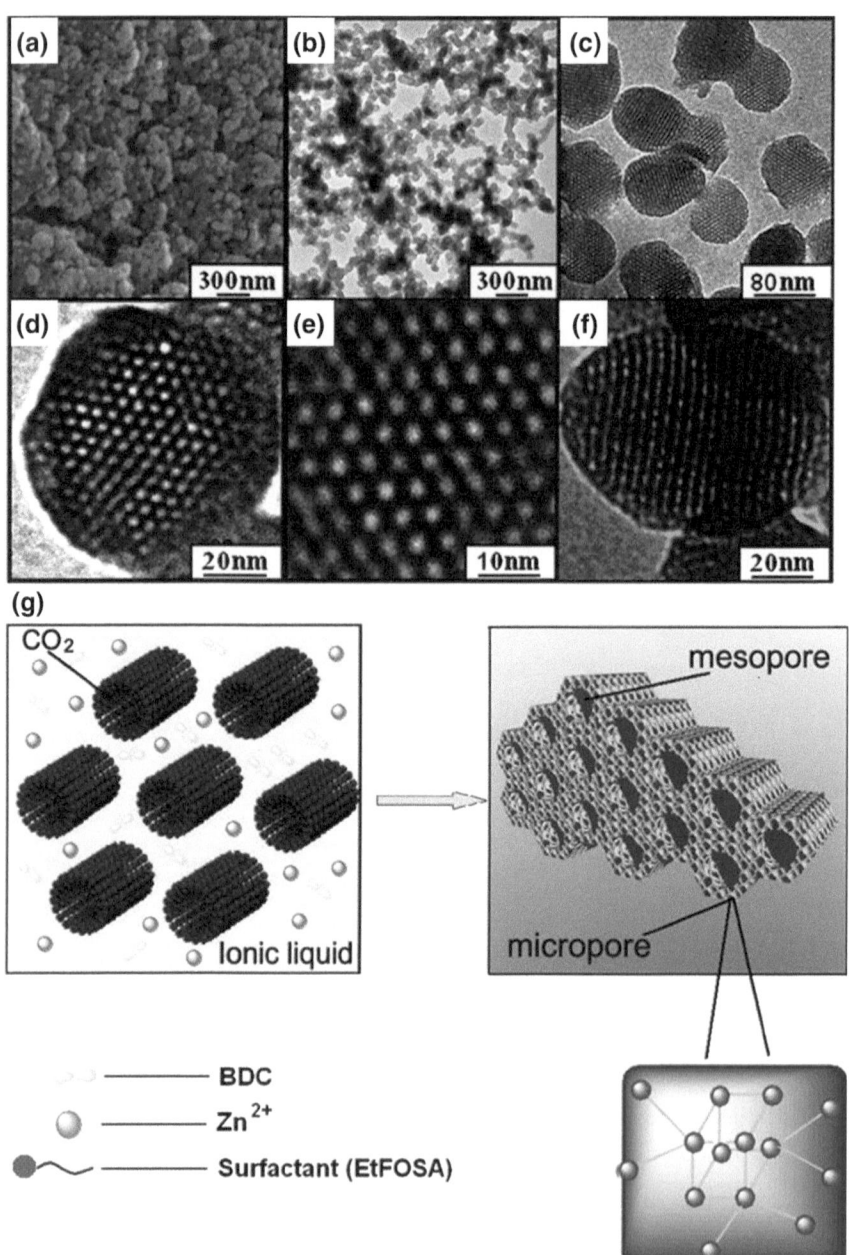

Fig. 3.10 a SEM and **b–f** TEM images of the mesoporous MOF. **g** Formation of the MOF in the surfactant/IL/CO$_2$ system. Reprinted with permission from Ref. [57]. Copyright 2010, Wiley-VCH

solvents and ion exchange, which could efficaciously prohibit the collapse of the mesoporous network.

Some general considerations are valid for all synthesis pathways described above in choosing the right precursors or synthetic conditions:

Firstly, the nature of the precursor species and the supermolecular surfactant has to allow a favorable interaction between the two components. This suggests that either a strong Coulomb interaction needs to be present, strong hydrogen-bonding forces act between the precursor molecules/related species and the sur-factant part of the composite to be formed, or hydrophilic/hydrophobic and electrostatic interaction prefer the formation of the mesostructures. If these inter-actions are not sufficiently strong, a high tendency for phase separation will domi-nate the reaction process, thus resulting in hybrid precipitation with poor porosity while the surfactant remaining in solution. Noticeably, this tendency will be espe-cially strong and rapid if the inorganic and organic precursors tend to form a stable crystalline structure with high lattice energy, or the hydrolysis rate of inorganic species is too fast to be controlled. It is thus not surprising that mesostructured silica-based materials are most easily accessible, which is due to that the tendency of silicon to form amorphous silica networks strongly and relatively low hydrol-ysis rate of siloxanes favors the formation of mesostructures. Indeed, a coopera-tive template system, comprising a surfactant (C_{16}TABr) and a chelating agent (citric acid), could favor the generation of a meso-MOF containing a hierarchical system of mesopores interconnected with microspores [59]. The surfactant mol-ecules formed micelles and the chelating agent bridged the MOF and the micelles, making self-assembly and crystal growth proceed under the direction of the cooperative template (Fig. 3.11). However, the resultant mesostructure is lack of long-range order.

Secondly, the inorganic and organic moieties should have a sufficiently high tendency to condense to an extended framework under the synthetic conditions. Otherwise, structures will collapse after the template removal. What should be

Fig. 3.11 Cooperative template-directed synthesis of mesoporous MOFs via self-assembly of metal ions and organic ligands. Reprinted with permission from Ref. [59]. Copyright 2012, American Chemical Society

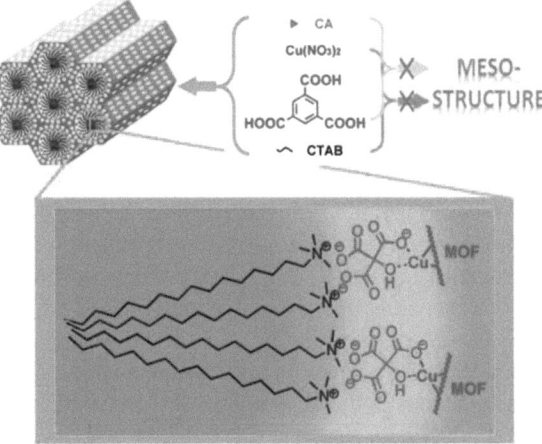

kept in mind is that template removal has to be possible without structural collapse. The most common method to remove the surfactant templates is mild extraction or ion exchange, and sometimes, high-temperature calcination is also used that may cause partial collapse of mesoporous networks and inevitably introduce some inorganic impurities. Extraction processes are best suited for nonionic surfactants. If ionic surfactants shall be extracted, the extraction typically has to be combined with an ion exchange because the surfactant also compensates framework charges, which may bring new functionalities such as high H^+ exchange capability.

Finally, if mesoporous materials shall be used under harsh conditions such as high-temperature reactions, problems can arise during reaction process. Crystallization of the hybrid framework represents an alternative way. However, the presence of the mesostructure and crystallinity of the walls are typically not compatible because crystalline materials can in most cases not accommodate the type of curvatures present in the mesostructures. Thus, as soon as crystallization of the walls to bulk occurs, the mesostructures deteriorate or collapse, as has been observed in many cases. Thus, further exploration including new synthesis strategies involving novel emerging techniques is worthy of tremendous endeavors.

3.2.3 Mesostructure Design

The adjustment of mesostructures is still a challenging task in the preparation of periodic mesoporous hybrid materials, while it is a key step in regulating their physicochemical properties. A lot of factors could cause the phase transformation of the mesopores, including interactions between the organic and inorganic species [60], the reaction temperature and time [61], the addition of some inorganic additives [62], the nature of surfactant that could be clarified using the molecular surfactant packing parameter [63], the molar ratios of the surfactant and inorganic precursor [64], and so forth. Herein, the molar ratios of the reactants were found to be critical for determining the final mesostructures of the hybrid materials, and different mesophases could be obtained by adjusting the amount of added reagents and surfactant. A synthesis condition map of periodic mesoporous titanium phosphonates with various phases has been explored (Fig. 3.12) [65]. The molar ratios of Ti/P should be fixed as 3:4 and 1:4 for hexagonal and cubic mesostructures, respectively, under the experimental conditions, while mixed phases with poor pore periodicity also existed at these two Ti/P ratios. It was easy to understand the Ti/P ratio in hexagonal mesophases because one Ti atom was coordinated to four P atoms, whereas one P was coordinated to three Ti atoms and was also connected to one C atom. The P species was superfluous in the cubic mesophases, which was probably due to the existence of some P atoms with a low-coordination state. By varying the molar ratios of $C_{16}TABr/Ti$, a general range for the synthesis of different mesophases was confirmed, namely in the region of $0.1 < C_{16}TABr/Ti < 0.4$ (Ti/P = 3:4) for a hexagonal phase, at $0.4 < C_{16}TABr/Ti < 1.9$ (Ti/P = 3:4, 1:4) for a mixed phase with poor pore

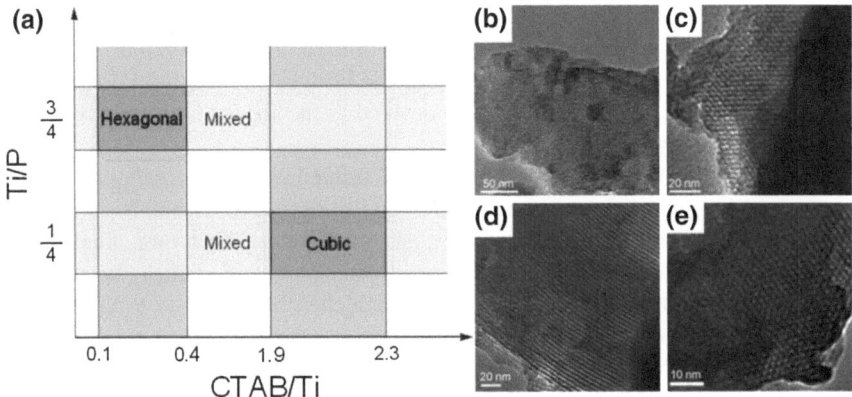

Fig. 3.12 **a** The synthesis condition map of periodic mesoporous titanium phosphonates, mainly considering the adding amounts of reagents and surfactant, TEM images of cubic (**b, c**) and hexagonal (**d, e**) mesoporous titanium phosphonate materials. Reprinted with permission from Ref. [65]. Copyright 2010, Wiley-VCH

regularity, and within the range of $1.9 < C_{16}TABr/Ti < 2.3$ (Ti/P = 1:4) for a cubic phase, which was in agreement with the previously reported molecular surfactant packing parameter theory that the hexagonal phase is formed at a low surfactant/ inorganic species ratio and the cubic phase formed at a high ratio [63]. Lamellar mesostructured aluminum organophosphonate with unique inorganic–organic hybrid network could be synthesized from the reactions of aluminum tri-isopropoxide with methylene disphosphonic acid with the assistance of alkyltrimethylammonium when the corresponding atomic ratios decreased to $Al/P/C_{16}TABr = 1:4:2$ [66]. The organic diphosphonic bridges were embedded in the integrated hybrid sheets with surfactant micelles inserted between the sheets. The removal of the surfactant would lead to the irreversible collapse of the lamellar phase, which signified that it had limited values from a practical applications point of view.

As to the mesoporous siliceous materials, the curvature of the mesostructures increases from lamellar via hexagonal to cubic phases, and the control of the mesostructure is commonly dependent on the hydrophilicity/hydrophobicity ratio and the molecular weight of the surfactants [67, 68]. The ease in tuning the surfactant composition via living polymerization paves the way for adjusting the mesophase. Initially, it is considered that lower hydrophilicity/hydrophobicity ratios lead to the formation of mesophases with small curvatures (e.g., lamellar), and high ratios are favorable for the generation of the ones with large curvatures (e.g., hexagonal and cubic) [69]. For example, the utilization of $EO_{80}PO_{30}EO_{80}$ with a high EO:PO ratio led to the preferential formation of cubic $Ia\bar{3}d$ phases and cage-type ones with $Fm\bar{3}m$ and $Pm\bar{3}m$ structures [70, 71]. In the cases of non-silica-based hybrid materials, the hexagonal mesophases are usually preferred in spite of the molecule structures and compositions. This may be due to the fact that the hybrid network condensed incompletely, and the arrangement of surfactant scaffolds containing inorganic species attached at the hydrophilic portions can make a difference from those of silica-based mesostructures [72]. Furthermore, the complex coordination chemistry between metal ions and organic bridges and the weak interactions among the organic components may be devoted to the formation of relatively stable hexagonal phases.

The final mesostructures are remarkably influenced by the rational control of organic–inorganic–organic interactions and cooperative assembly of precursor molecules and their related species and surfactants. Therefore, final mesostructures depend on the surfactant liquid–crystal phases. Significantly, different types of surfactants possess distinct critical micelle concentration (CMC), while a surfactant with low CMC value is an important criterion toward enhancing the regularity of mesostructures. Surfactants with high CMC usually lead to cubic mesophases. If the CMC values further increase, it is difficult to produce periodic mesostructures. Different synthesis pathway has its own advantages and disadvantages. EISA method can fit wide preparation conditions owing to the alleviated hydrolysis speed of metallic precursors, though this strategy needs relatively rigorous conditions including temperature and humidity. Hydrothermal autoclaving is a quick and efficient route, though energy consumption is required for the high temperature and pressure. Ionic liquid is high cost and not suitable for broad application. One should choose the most suitable pathway depending on the practical situation. Exploring low cost, environmentally friendly, and reproducible method is still an urgent requirement.

On the other hand, mesoporous materials can be divided into ordered and disordered structures from the viewpoint of pore/channel packing regularity. The syntheses of disordered wormhole-like mesoporous non-siliceous hybrid materials templated by supramolecular surfactants have also contributed a great deal to the exploiting of the organization principles in the surfactant-assisted strategy. Disordered mesostructures have no unit cell, symmetry, or space group. Although the resultant mesostructures are disordered, uniformly sized mesopores, high surface area, and easy modulation can usually be achieved, offering them potential in catalysis, adsorption, separation, and immobilization. Due to the complexity, it is still difficult to pass a definitive verdict as to which kind of mesostructures, disordered or ordered, is more beneficial in applications. The exploration of the territory of preparing mesoporous phosphonate-based hybrids is not as mature as the silicas and carbonaceous materials owing to the complex interconnected factors, coordination, and solgel chemistry. Thus, further investigation should be of significance and justifies investment.

3.2.4 Pore Size Control

The control of the pore size of mesoporous materials is a vital issue, as it is directly linked to their application. Pore sizes mainly depend on the hydrophobic volumes of the template molecules. With this core idea in mind, the intentional selection of surfactants with various lengths of hydrophobic chains can

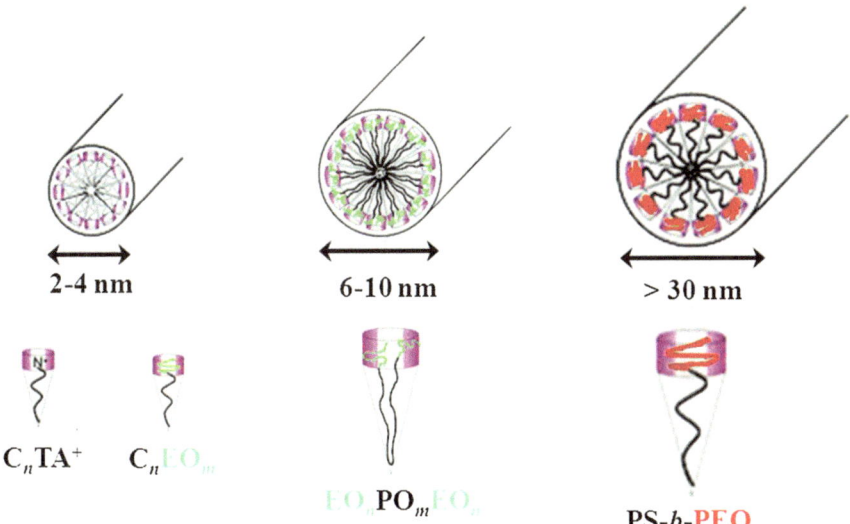

2-4 nm   **6-10 nm**   **> 30 nm**

C_nTA^+ C_nEO_m

$EO_nPO_mEO_n$ PS-*b*-PEO

Fig. 3.13 Adjustment of the mesoporous size based on surfactant molecules with different hydrophobic volumes

make it possible to control the pore size. Using soft templates with low molecular weights (Fig. 3.13), such as $C_{16}TABr$ and oligomeric surfactants (e.g., $C_{16}EO_{10}$ and $C_{16}EO_{20}$), the pore diameters are usually distributed in the range of 2–4 nm [46–48, 52, 55, 56]. In general, the block–copolymer micelles are larger than those aggregated by low molecular weight surfactants. For instance, the pore width could be largely expanded to 6–10 nm if triblock copolymers ($EO_{80}PO_{30}EO_{80}$, $EO_{106}PO_{70}EO_{106}$, and $EO_{20}PO_{70}EO_{20}$) were employed [48, 73, 74]. On the other side, diblock–copolymers always direct larger pore sizes compared to triblock–copolymers with similar molecular weights or PPO chains because the latter tends to bending aggregation. High molecular weight block–copolymers are of great interest owing to the relatively large mesopores in the resultant mesoporous materials. Colloidal templating of polystyrene-block-poly(oxyethylene) (PS-*b*-PEO) was newly developed for the fabrication of porous aluminum phosphonates with large spherical pores [75]. PS-*b*-PEO was dissolved in a mixture of THF and ethanol in the presence of water to form spherical aggregates and then mixed with precursor solutions before preparation. Then, the mixed solutions were spray-dried at different temperatures, from 110 to 230 °C, and calcined at an appropriate temperature to eliminate the surfactants. Interestingly, the size of the spherical PS-*b*-PEO aggregates (the number of PS-*b*-PEO molecules in the aggregates) was variable on the basis of the amount of water present. Therefore, the pore diameter could be controlled from large mesopores (30 nm) to macropores (200 nm), as it was related to the size of the colloidal PS-*b*-PEO template.

The addition of organic swelling agents is another significant way to expand the pore sizes. The hydrophobic organic species can be solubilized inside the hydrophobic regions of the surfactant micelles based on hydrophobicity–hydrophobicity interactions, leading to micelle swelling. With $C_{16}TABr$ acting as a structure-directing agent and 1,3,5-trimethylbenzene (TMB) as an auxiliary one, Qiu et al. [76] prepared a series of hierarchically porous HKUST-1 with adjustable interconnecting micropores and mesopores by the self-assembly of framework–building blocks in the presence of surfactant micelles. The synthesized mesostructured MOFs possessed a mesopore system with diameters tunable from 3.8 to 31.0 nm, which depended on the synthetic conditions. Additionally, the mesoporous walls were constructed by a crystalline microporous network containing a 3D system of channels with a pore diameter of 0.82 nm, which was confirmed by the XRD, N_2 sorption, and TEM analysis. The surface area of the MOFs decreased remarkably from 1,124 to 579 $m^2 \ g^{-1}$ with the increase of the TMB/$C_{16}TABr$ molar ratio from 0 to 0.50. It is noteworthy that the enlargement of the pore sizes usually implicates the sacrifice of the specific surface area.

The effect of the nature of surfactant species and the addition of organic swelling agents on the pore size control were systematically studied [77]. Spherical aluminum phosphonate particles were obtained using Pluronic F127 with the formation of larger mesopores (10.5 nm) than those architectured using Brij58 (4.1 nm) and Pluronic F68 (6.3 nm). Further expansion of the mesopores from 10.5 to 15 and 20 nm was achieved by the addition of typical aromatic compounds 1,3,5-TMB and 1,3,5-tri-isopropylbenzene as organic swelling agents, respectively. The boiling points of the aromatic compounds were quite important for the successful fabrication of high-quality spherical particles of the ordered mesoporous aluminum phosphonates. In addition, when aromatic compounds are regarded as model molecules, aerosol-assisted fabrication in the presence of designed and/or synthetic organic compounds such as fine chemicals is quite potential for the production of spherical supports with ordered mesopores.

These phenomenons have indeed given us a hint. Other substances that can be dissolved in the micelle cores may also expand pore diameters. Binary surfactant systems can result in products with tunable pore sizes and bimodal or trimodal pores. This has been testified in mesoporous siliceous materials. For example, blending two quaternary cationic surfactants with different carbon chains together (e.g., $C_{12}TABr$, $C_{16}TABr$, $C_{16}TABr$, and $C_{22}TABr$) can change the pore sizes of MCM-41 mesostructures to intermediate values between that templated by a single surfactant [78]. With respect to mesoporous non-siliceous materials, a step further involving binary and even ternary surfactant system is full of interest to be explored.

Additionally, the reaction medium of the synthetic systems, as well as the crystallization time, is also inevitable to affect the pore size of the resultant mesostructured hybrids. A series of amorphous porous zirconium phosphonate materials constructed from HEDP, having tunable from micropore to mesopore sizes, were hydrothermally synthesized in a $C_{16}TABr$–H_2O–ethanol ternary system [79]. The as-synthesized materials were mesostructured and could be transformed into (super-)microporous hybrid solids after surfactant-extracted process. It was observed that ethanol played a role as cosurfactant in assisting the formation of zirconium organophosphonate mesostructures, by achieving the charge density matching between the surfactant and hybrid nanoparticles. By varying the hydrothermal time from 1 to 72 h at 110 °C, the pore sizes of the obtained zirconium phosphonates could be efficiently tuned from micropore (0.87 nm) to mesopore (2.5 nm) range, and their micropore-specific surface areas ranged from 116 to 509 m^2 g^{-1} with the pore volumes in the range of 0.11–0.35 cm^3 g^{-1}. Considering the acidity and tunable pore sizes, the prepared porous zirconium phosphonates may find their potential applications in adsorption, shape-selective heterogeneous catalysis, ion exchange, and proton conduction.

3.2.5 Improving the Crystallization of Pore Walls

A high degree of crystallinity of the hybrid structures is quite significant for improving the properties and thus the applicability of these materials in many fields. Attempts to crystallize the materials through heat treatment always resulted in the collapse of the periodic mesoporous structures and the deterioration of the hybrid frameworks, which could impede their extended applications [80]. Revealingly, using organosilane instead of inorganic precursors, organic–inorganic hybrid periodic mesoporous benzene–silica was synthesized with a crystalline wall structure, formed from the structure-directing interactions between the benzene–silica precursor molecules and between the precursor molecules and the surfactants [81], which has supplied scientific researchers with an alternative method to prepare crystalline mesostructures of hybrid materials.

Kimura et al. [82] reported the preparation of lamellar mesostructured aluminum phosphonates with a crystalline hybrid framework from the reaction of aluminum triisopropoxide with methylene diphosphonic acid in the presence of alkyltrimethylammonium C_nTMA ($n \geq 14$) surfactants. This was only under restricted conditions, and the crystal structure of the hybrid framework has not been defined yet because of the ill-resolved diffraction peaks. In the study, methylene groups can be embedded in the sheets (integrated hybrid frameworks) of the lamellar AOP-1 by a surfactant-assisted strategy. Al atoms are six-coordinated, connected to oxygen atoms only, and bonded to P atoms through oxygen atoms, and water molecules are ligated to the Al atoms. However, the ethylene groups are embedded in the hybrid framework, but the framework is not constructed from Al, P, and the organic group only. P atoms are bonded to AlO_6 units, which is a common structure for the reported crystalline AOPs. The six-coordinated Al atoms are surrounded by four oxygen atoms (two Al–O–P bonds and two water ligands) and two F atoms, meaning that the framework is not pure. On the other side, such hybrid mesostructures are unstable, and the surfactants could not be removed without the collapse of the mesostructure.

Microwave-assisted synthesis has been shown to be a facile and moderate approach for promoting the crystallization of porous materials at a relatively low temperature [83, 84]. Ma et al. [73] reported the successful preparation of ordered, hexagonal, mesoporous metal (Ti, Zr, V, and Al) phosphonate materials with microporous crystalline walls using a microwave-assisted procedure in the presence of triblock copolymer F127 as a template. Corresponding metal chlorides and ethylene diamine tetra(methylene phosphonic acid) were chosen as the inorganic precursors and the coupling molecule, respectively. The most important benefit of applying microwave irradiation here is probably the fast dissolution of the gel and the simultaneous abundant nucleation in the synthetic mixture, caused by the rapid heating and efficient heat transfer using this technique, which consequently results in the fast crystallization and high crystallinity [85, 86]. During the microwave treatment, the small nascent metal phosphonate crystals are formed. However, the microwave treatment seems hardly contributive to the formation of ordered mesopores. Thus, the periodic assembly of phosphonate crystals along the surfactant F127 micelles to form a hexagonal mesophase could only be realized after a hydrothermal aging process, which helps the further crystallization of small crystals. The prepared metal phosphonates possessed a hierarchical porous structure with pore sizes of 7.1–7.5 nm for mesopores and 1.3–1.7 nm for micropores, respectively, and were thermally stable up to approximately 450 °C, with the pore structure and hybrid framework well preserved. The ordered hexagonal mesopores and one-dimensional pore channels could be confirmed using TEM images, and crystal lattice fringes could be observed in the magnified image of the pore walls (Fig. 3.14). The crystalline phase of these phosphonate-based hybrids could be attributable to the corresponding metal phosphonates crystals formed by the extensive coordination of phosphonic claw groups with metal ions rather than tiny metal oxide particles. The phosphonate groups are homogenously incorporated into the hybrid framework of the obtained materials.

The pivotal factor to obtain well-defined mesoporosity and fine crystallization is to slow down the coordination rates between the metal centers and organic linkers so as to allow the assembly of nanosized building blocks and surfactant micelles. Acetic acid can chelate many metal ions, such as Cu^{2+}, to form a derivative of metal–acetate bidentate bridging [87]. Namely, the acetic acid could compete with the carboxylate linkers to coordinate with metal ions and influence the deprotonation of the linker as well. Under the synergic effect of both factors, phase segregation was limited thus to fit the liquid–crystal templating mechanism. N_2 adsorption–desorption, TEM, and XRD indicated the generation of well-defined mesopore channels within the microporous copper carboxylates, which presented high crystallinity assigned to the HKUST-1. However, the mesostructure possessed no long-range order.

The positively charged surfactants (S^0H^+) and cationic inorganic species (I^+) are assembled together by a combination of electrostatic, hydrogen-bonding, and van der Waals interactions (S^0H^+) X^-I^+ ($X^- = Cl^-$, NO_3^-, $HySO_4^{2-y}$, etc.). If X^- was substituted by disulfonate anions, an analogous mechanism was proposed to be accomplished to form a series of highly ordered mesoporous metal sulfonates [88]. The coordination expansion based on X^-I^+ could form pillared-layered metal disulfonates in the mesoporous wall (Fig. 3.15, a kind of cadmium disulfonate crystal is taken as a representative). In a typical synthesis procedure

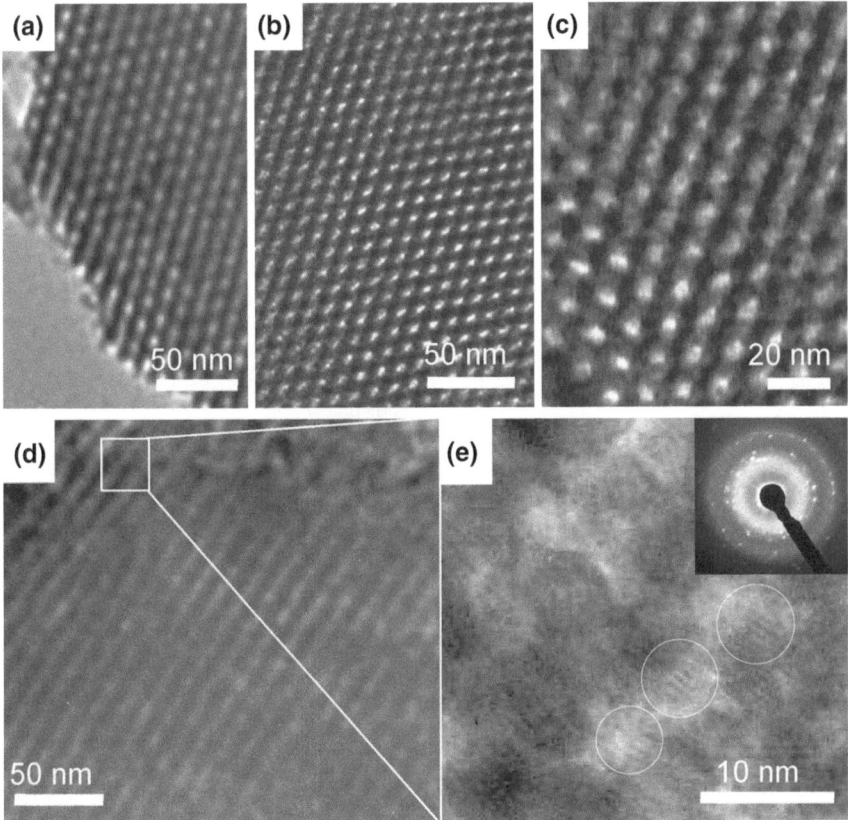

Fig. 3.14 TEM images (**a–e**) and SAED pattern (*inset* of **e**) of the semicrystalline titanium phosphonate material. The circles in **e** highlight the intercrystalline micropores. Reprinted with permission from Ref. [73]. Copyright 2011, Wiley-VCH

for mesoporous metal–organic framework material, MMOF-1, $Cd(NO_3)_2$, and 1,5-naphthalenedisulfonic acid (1,5-nds) were used as inorganic and organic precursors, respectively, and nonionic triblock copolymer F127 was used as the soft template in the acidic system. The controlled release of metal ions by crown ether 1,10-diaza-18-crown-6 (NC) is necessary to slow down the coordination rate between the inorganic and organic species and thus finally to retard nucleation kinetics and crystal growth of the MOFs around the micelles, resulting in the final formation of well-structured hexagonal mesoporosity by the surfactant F127-induced self-assembly process, with the mesopore walls constructed by crystalline metal disulfonate $[Cu(1,5\text{-}nds)(H_2O)_2]_n$. Other common ligands including ethylenediamine, ethylene diamine tetraacetic acid, nitrilotriacetic acid, 2,2′-dipyridyl, and ethylene diamine tetra(methylene phosphonic acid) were also tried in order to replace crown ether, but no ordered mesopores were observed in the dense crystalline products, and in some cases, impure phases related to the coordination of

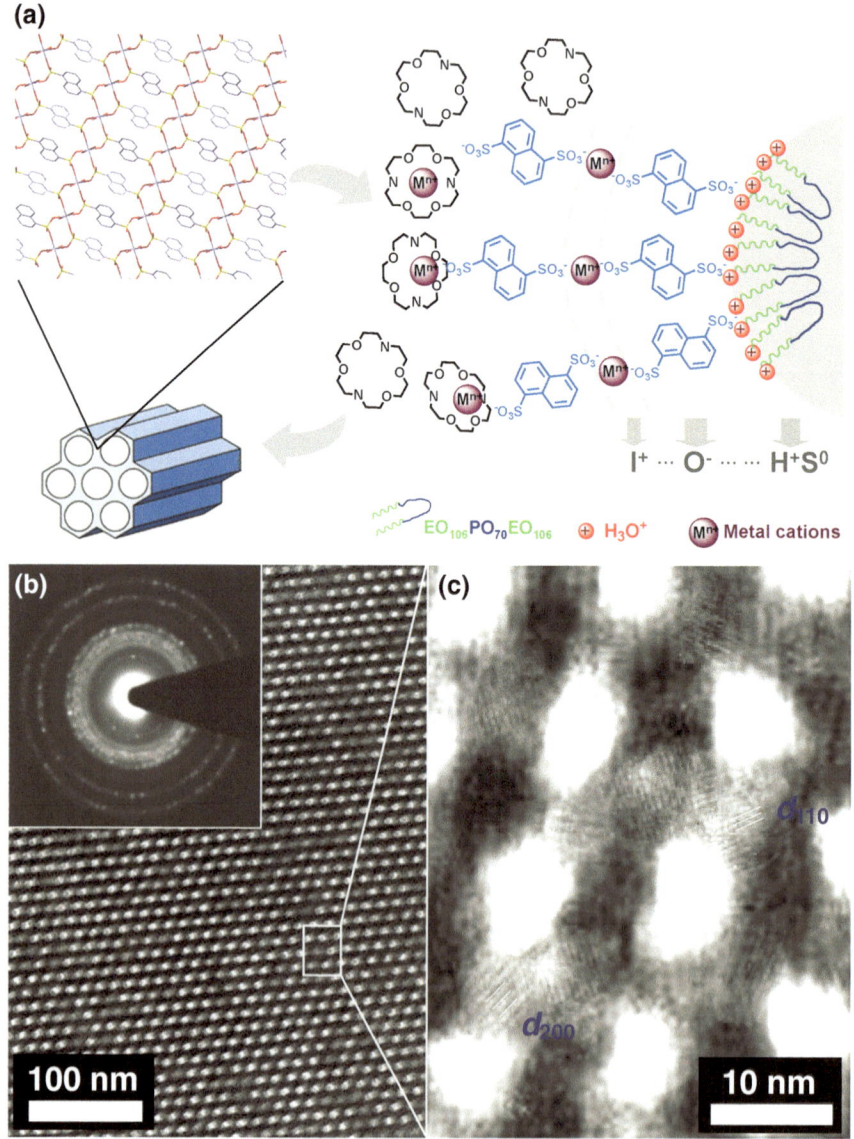

Fig. 3.15 Schematic model for the formation of mesoporous MOFs, where the coordination numbers in the right figure do not reflect the real situation (**a**). TEM images and electron diffraction pattern (**b, c**). Reprinted with permission from Ref. [88]. Copyright 2012, American Chemical Society

newly added ligands with metal cations also appeared. This was probably due to the hydrophobic surface of NC being able to effectively isolate the innerly embedded metal cations from the sulfonate anions during the hydrothermal process. This strategy could be extended to prepare other mesoporous MOFs. Using 1,5-nds, mesoporous MOFs constructed by metal disulfonate $[La(1,5\text{-}nds)1.5(H_2O)_5]_n$ and $[Cu(1,5\text{-}nds)(H_2O)_4]_n$ were successfully prepared, named as MMOF-2 and MMOF-3, respectively. Mesoporous MOFs consisting of layered $Sr[C_2H_4(SO_3)_2]$ were obtained using ethanedisulfonic acid, named as MMOF-4. It is expected that the crystalline mesoporous frameworks can be constructed from many other types of easily crystallized materials by controlled release of metal ions, leading to enhanced functionality. This soft-templating assembly route should be generally transferable to other mesoporous MOFs with an appropriate choice of functional groups.

The controlled synthesis of mesoporous non-siliceous organic–inorganic hybrids includes the adjustment of pore sizes, mesophase symmetry, crystallinity of the pore walls, and micro-/macroscopic morphologies. Various mature and burgeoning technologies can be employed to attain the desired targets. However, the achievement of well-structured mesophases and a high crystalline degree at the same time are still a contradiction. The effective control of the mesophase is still challenging, but especially significant in the areas of adsorption, separation, and catalysis.

References

1. G. Alberti, U. Costantino, F. Marmottini, R. Vivani, P. Zapelli, Preparation of a covalently pillared alpha-zirconium phosphite-diphosphonate with a high degree of interlayer porosity. Micropor. Mesopor. Mater. **21**, 297–304 (1998)
2. G. Alberti, R. Vivani, F. Marmottini, P. Zappelli, Microporous solids based on pillared metal(IV) phosphates and phosphonates. J. Porous Mater. **5**, 205–220 (1998)
3. A. Clearfield, Organically pillared micro- and mesoporous materials. Chem. Mater. **10**, 2801–2810 (1998)
4. S. Konar, J. Zon, A.V. Prosvirin, K.R. Dunbar, A. Clearfield, Inorg. Chem. **26**, 5229–5236 (2007)
5. G.K.H. Shimizu, R. Vaidhyanathan, J.M. Taylor, Phosphonate and sulfonate metal organic frameworks. Chem. Soc. Rev. **38**, 1430–1449 (2009)
6. F. Gándara, A. Garca-Corteís, C. Cascales, B. Gómez-Lor, E. Gutiérrez-Puebla, M. Iglesias, A. Monge, N. Snejko, Rare earth arenedisulfonate metal-organic frameworks: an approach toward polyhedral diversity and variety of functional compounds. Inorg. Chem. **46**, 3475–3484 (2007)
7. D.J. Hoffart, S.A. Dalrymple, G.K.H. Shimizu, Structural constraints in the design of silver sulfonate coordination networks: three new polysulfonate open frameworks. Inorg. Chem. **44**, 8868–8875 (2005)
8. L. Ma, J.M. Falkowski, C. Abney, W. Lin, A series of isoreticular chiral metal-organic frameworks as a tunable platform for asymmetric catalysis. Nature Chem. **2**, 838–846 (2010)
9. O.M. Yaghi, H. Li, C. Davis, D. Richardson, T.L. Groy, Synthetic strategies, structure patterns, and emerging properties in the chemistry of modular porous solids. Acc. Chem. Res. **31**, 474–484 (1998)
10. X.S. Wang, S.Q. Ma, D.F. Sun, S. Parkin, H.C. Zhou, A mesoporous metal-organic framework with permanent porosity. J. Am. Chem. Soc. **128**, 16474–16475 (2006)
11. Y. Yan, S.H. Yang, A.J. Blake, W. Lewis, E. Poirier, S.A. Barnett, N.R. Champness, M. Schröder, A mesoporous metal-organic framework constructed from a nanosized C-3-symmetric linker and [Cu-24(isophthalate)(24)] cuboctahedra. Chem. Commun. **47**, 9995–9997 (2011)

12. H. Deng, S. Grunder, K.E. Cordova, C. Valente, H. Furukawa, M. Hmadeh, F. Gándara, A.C. Whalley, Z. Liu, S. Asahina, H. Kazumori, M. O'Keeffe, O. Terasaki, J.F. Stoddart, O.M. Yaghi, Large-pore apertures in a series of metal-organic frameworks. Science **336**, 1018–1023 (2012)

13. M. Vasylyev, E.J. Wachtel, R. Popovitz-Biro, R. Neumann, Titanium phosphonate porous materials constructed from dendritic tetraphosphonates. Chem. Eur. J. **12**, 3507–3514 (2006)

14. M. Vasylyev, R. Neumann, Preparation, characterization, and catalytic aerobic oxidation by a vanadium phosphonate mesoporous material constructed from a dendritic tetraphosphonate. Chem. Mater. **18**, 2781–2783 (2006)

15. A.O. Ibrahim, Y. Zhou, F. Jiang, L. Chen, X. Li, W. Xu, O.O.E. Onawumi, O.A. Odunola, M. Hong, An unusual (10,3)-d MOF material with nanoscale helical cavities and multifunctionality. Eur. J. Inorg. Chem. **32**, 5000–5005 (2011)

16. T.Y. Ma, X.J. Zhang, Z.Y. Yuan, Hierarchically meso-/macroporous titanium tetraphosphonate materials: synthesis, photocatalytic activity and heavy metal ion adsorption. Micropor. Mesopor. Mater. **123**, 234–242 (2009)

17. P. Schmidt-Winkel, W.W. Lukens, D.Y. Zhao, P.D. Yang, B.F. Chmelka, G.D. Stucky, Mesocellular siliceous foams with uniformly sized cells and windows. J. Am. Chem. Soc. **121**, 254–255 (1999)

18. X.J. Zhang, T.Y. Ma, Z.Y. Yuan, Nanostructured titania-diphosphonate hybrid materials with a porous hierarchy. Eur. J. Inorg. Chem. **2008**(17), 2721–2726 (2008)

19. X.Y. Yang, A. Léonard, A. Lemaire, G. Tian, B.L. Su, Self-formation phenomenon to hierarchically structured porous materials: design, synthesis, formation mechanism and applications. Chem. Commun. **47**, 2763–2786 (2011)

20. Z.Y. Yuan, B.L. Su, Insights into hierarchically meso-macroporous structured materials. J. Mater. Chem. **16**, 663–677 (2006)

21. T.Y. Ma, X.J. Zhang, Z.Y. Yuan, Hierarchical meso-/macroporous aluminum phosphonate hybrid materials as multifunctional adsorbents. J. Phys. Chem. C **113**, 12854–12862 (2009)

22. C. Liu, T. Li, N.L. Rosi, Strain-promoted "click" modification of a mesoporous metal-organic framework. J. Am. Chem. Soc. **134**, 18886–18888 (2012)

23. S. Yang, X. Lin, W. Lewis, M. Suyetin, E. Bichoutskaia, J.E. Parker, C.C. Tang, D.R. Allan, P.J. Rizkallah, P. Hubberstey, N.R. Champness, K.M. Thomas, A.J. Blake, M. Schröder, A partially interpenetrated metal-organic framework for selective hysteretic sorption of carbon dioxide. Nat. Mater. **11**, 710–716 (2012)

24. Y.F. Yue, Z.A. Qiao, P.F. Fulvio, A.J. Binder, C.C. Tian, J.H. Chen, K.M. Nelson, X. Zhu, S. Dai, Template-free synthesis of hierarchical porous metal-organic frameworks. J. Am. Chem. Soc. **135**, 9572–9575 (2013)

25. T. Tsuruoka, S. Furukawa, Y. Takashima, K. Yoshida, S. Isoda, S. Kitagawa, Nanoporous nanorods fabricated by coordination modulation and oriented attachment growth. Angew. Chem. Int. Ed. **48**, 4739–4743 (2009)

26. A.R. Hirst, B. Escuder, J.F. Miravet, D.K. Smith, High-tech applications of self-assembling supramolecular nanostructured gel-phase materials: from regenerative medicine to electronic devices. Angew. Chem. Int. Ed. **47**, 8002–8018 (2008)

27. Q. Wei, S.L. James, A metal-organic gel used as a template for a porous organic polymer. Chem. Commun. **41**, 1555–1556 (2005)

28. Y.R. Liu, L. He, J. Zhang, X. Wang, C.Y. Su, Evolution of spherical assemblies to fibrous networked Pd(II) metallogels from a pyridine-based tripodal ligand and their catalytic property. Chem. Mater. **21**, 557–563 (2009)

29. L. Li, S.L. Xiang, S.Q. Cao, J.Y. Zhang, G.F. Ouyang, L.P. Chen, C.Y. Su, A synthetic route to ultralight hierarchically micro/mesoporous Al(III)-carboxylate metal-organic aerogels. Nat. Commun. **4**, 1774 (2013)

30. D. Zacher, R. Schmid, C. Woll, R.A. Fischer, Surface chemistry of metal-organic frameworks at the liquid-solid interface. Angew. Chem. Int. Ed. **50**, 176–199 (2011)

31. J. Cravillon, C.A. Schröder, R. Nayuk, J. Gummel, M. Wiebcke, Fast nucleation and growth of ZIF-8 nanocrystals monitored by time-resolved in situ small-angle and wide-angle X-ray scattering. Angew. Chem. Int. Ed. **50**, 8067–8071 (2011)

32. S. Cao, G. Gody, W. Zhao, S. Perrier, X.Y. Peng, C. Ducati, D.Y. Zhao, A.K. Cheetham, Hierarchical bicontinuous porosity in metal-organic frameworks templated from functional block co-oligomer micelles. Chem. Sci. **4**, 3573–3577 (2013)
33. Y.P. Zhu, Y.L. Liu, T.Z. Ren, Z.Y. Yuan, Mesoporous nickel phosphate/phosphonate hybrid microspheres with excellent performance for adsorption and catalysis. RSC Adv. **4**, 16018–16021 (2014)
34. W. Li, D.Y. Zhao, An overview of the synthesis of ordered mesoporous materials. Chem. Commun. **49**, 943–946 (2013)
35. T.Y. Ma, L. Liu, Z.Y. Yuan, Direct synthesis of ordered mesoporous carbons. Chem. Soc. Rev. **42**, 3977–4003 (2013)
36. A. Fischereder, M.L. Martinez-Ricci, A. Wolosiuk, W. Haas, F. Hofer, G. Trimmel, G.J.A.A. Soler-Illia, Mesoporous ZnS thin films prepared by a nanocasting route. Chem. Mater. **24**, 1837–1845 (2012)
37. P. Yang, D. Zhao, D.I. Margolese, B.F. Chmelka, G.D. Stucky, Block copolymer templating syntheses of mesoporous metal oxides with large ordering lengths and semicrystalline framework. Chem. Mater. **11**, 2813–2826 (1999)
38. J.S. Yu, S. Kang, S.B. Yoon, G. Chai, Fabrication of ordered uniform porous carbon networks and their application to a catalyst supporter. J. Am. Chem. Soc. **124**, 9382–9383 (2002)
39. T.F. Baumann, J.H. Satcher, Homogeneous incorporation of metal nanoparticles into ordered macroporous carbons. Chem. Mater. **15**, 3745–3747 (2003)
40. Q.D. Nghiem, D.P. Kim, Direct preparation of high surface area mesoporous SiC-based ceramic by pyrolysis of a self-assembled polycarbosilane-block-polystyrene diblock copolymer. Chem. Mater. **20**, 3735–3739 (2008)
41. S. Che, A.E. Garcia-Bennett, T. Yokoi, K. Sakamoto, H. Kunieda, Q. Terasaki, T. Tatsumi, A novel anionic surfactant templating route for synthesizing mesoporous silica with unique structure. Nat. Mater. **2**, 801–805 (2003)
42. B.Z. Tian, X.Y. Liu, B. Tu, C. Yu, J. Fan, L.M. Wang, S.H. Xie, G.D. Stucky, D.Y. Zhao, Self-adjusted synthesis of ordered stable mesoporous minerals by acid-base pairs. Nat. Mater. **2**, 159–163 (2003)
43. S.A. Davis, M. Breulmann, K.H. Rhodes, B. Zhang, S. Mann, Template-directed assembly using nanoparticle building blocks: a nanotectonic approach to organized materials. Chem. Mater. **13**, 3218–3226 (2001)
44. X. Roy, L.K. Thompson, N. Coombs, M.J. MacLachlan, Mesostructured Prussian blue analogues. Angew. Chem. Int. Ed. **47**, 511–514 (2008)
45. Y.T. Li, D.J. Zhang, Y.N. Guo, B.Y. Guan, D.H. Tang, Y.L. Liu, Q.S. Huo, Design and synthesis of novel mesostructured metal-organic frameworks templated by cationic surfactants via cooperative self-organization. Chem. Commun. **47**, 7809–7811 (2011)
46. T. Kimura, Synthesis of mesostructured and mesoporous aluminum organophosphonates prepared by using diphosphonic acids with alkylene groups. Chem. Mater. **15**, 3742–3744 (2003)
47. T. Kimura, Synthesis of mesostructured and mesoporous aluminum organophosphonates prepared by using diphosphonic acids with alkylene groups. Chem. Mater. **17**, 337–344 (2005)
48. T. Kimura, Oligomeric surfactant and triblock copolymer syntheses of aluminum organophosphonates with ordered mesoporous structures. Chem. Mater. **17**, 5521–5528 (2005)
49. J.E. Haskouri, C. Guillem, J. Latorre, A. Beltrán, D. Beltrán, P. Amorós, $S^{+}I^{-}$ ionic formation mechanism to new mesoporous aluminum phosphonates and diphosphonates. Chem. Mater. **16**, 4359–4372 (2004)
50. J.E. Haskouri, C. Guillem, J. Latorre, A. Beltrán, D. Beltrán, P. Amorós, The first pure mesoporous aluminium phosphonates and diphosphonates-new hybrid porous materials. Eur. J. Inorg. Chem. **9**, 1804–1807 (2004)
51. S. Cabrera, J.E. Haskouri, C. Guillem, J. Latorre, A. Beltrán, D. Beltrán, M.D. Marcos, P. Amorós, Generalised syntheses of ordered mesoporous oxides: the atrane route. Solid State Sci. **2**, 405–420 (2000)

52. T.Y. Ma, X.Z. Lin, Z.Y. Yuan, Periodic mesoporous titanium phosphonate hybrid materials. J. Mater. Chem. **20**, 7406–7415 (2010)
53. G.J.A.A. Soler-Illia, C. Sanchez, Interactions between poly(ethylene oxide)-based surfactants and transition metal alkoxides: their role in the templated construction of mesostructured hybrid organic-inorganic composites. New J. Chem. **24**, 493–499 (2000)
54. G.J.A.A. Soler-Illia, E. Scolan, A. Louis, P.A. Albouy, C. Sanchez, Design of meso-structured titanium oxo based hybrid organic-inorganic networks. New J. Chem. **25**, 156–165 (2001)
55. T.Y. Ma, Z.Y. Yuan, Functionalized periodic mesoporous titanium phosphonate monoliths with large ion exchange capacity. Chem. Commun. **46**, 2325–2327 (2010)
56. T.Y. Ma, Z.Y. Yuan, Periodic mesoporous titanium phosphonate spheres for high dispersion of CuO nanoparticles. Dalton Trans. **39**, 9570–9578 (2010)
57. Y.J. Zhao, J.L. Zhang, B.X. Han, J.L. Song, J.S. Li, Q. Wang, Metal-organic framework nanospheres with well-ordered mesopores synthesized in an ionic liquid/CO_2/surfactant system. Angew. Chem. Int. Ed. **50**, 636–639 (2011)
58. J.H. Liu, S.Q. Cheng, J.L. Zhang, X.Y. Feng, X.G. Fu, B.X. Han, Reverse micelles in carbon dioxide with ionic-liquid domains. Angew. Chem. Int. Ed. **46**, 3313–3315 (2007)
59. L.B. Sun, J.R. Li, J. Park, H.C. Zhou, Cooperative template-directed assembly of mesoporous metal-organic frameworks. J. Am. Chem. Soc. **134**, 126–129 (2012)
60. D.Y. Zhao, Q.S. Huo, J.L. Feng, B.F. Chmelka, G.D. Stucky, Nonionic triblock and star diblock copolymer and oligomeric surfactant syntheses of highly ordered, hydrothermally stable, mesoporous silica structures. J. Am. Chem. Soc. **120**, 6024–6036 (1998)
61. D.Y. Zhao, J.L. Feng, Q.S. Huo, N. Melosh, G.H. Fredrickson, B.F. Chmelka, G.D. Stucky, Triblock copolymer syntheses of mesoporous silica with periodic 50 to 300 angstrom pores. Science **279**, 548–552 (1998)
62. W.J. Kim, J.C. Yoo, D.T. Hayhurst, Synthesis of MCM-48 via phase transformation with direct addition of NaF and enhancement of hydrothermal stability by post-treatment in NaF solution. Micropor. Mesopor. Mater. **49**, 125–137 (2001)
63. Q.S. Huo, D.I. Margolese, G.D. Stucky, Surfactant control of phases in the synthesis of mesoporous silica-based materials. Chem. Mater. **8**, 1147–1160 (1996)
64. J.C. Vartuli, K.D. Schmitt, C.T. Kresge, W.J. Roth, M.E. Leonowicz, S.B. McCullen, S.D. Hellring, J.S. Beck, J.L. Schlenker, D.H. Olson, E.W. Sheppard, Effect of surfactant/silica molar ratios on the formation of mesoporous molecular sieves: inorganic mimicry of surfactant liquid–crystal phases and mechanistic implications. Chem. Mater. **6**, 2317–2326 (1994)
65. T.Y. Ma, X.Z. Lin, Z.Y. Yuan, Cubic mesoporous titanium phosphonates with multifunctionality. Chem. Eur. J. **16**, 8487–8494 (2010)
66. T. Kimura, D. Nakashima, N. Miyamoto, Lamellar mesostructured aluminum organophosphonate with unique crystalline framework. Chem. Lett. **38**, 916–917 (2009)
67. D. Zhao, Q. Huo, J. Feng, J. Kim, Y. Han, G.D. Stucky, Novel mesoporous silicates with two-dimensional mesostructure direction using rigid bolaform surfactants. Chem. Mater. **11**, 2668–2672 (1999)
68. Y.H. Deng, J. Wei, Z.K. Sun, D.Y. Zhao, Large-pore ordered mesoporous materials templated from non-Pluronic amphiphilic block copolymers. Chem. Soc. Rev. **42**, 4054–4070 (2013)
69. B.C. Garcia, M. Kamperman, R. Ulrich, A. Jain, S.M. Gruner, U. Wiesner, Morphology diagram of a diblock copolymer-aluminosilicate nanoparticle system. Chem. Mater. **21**, 5397–5405 (2009)
70. S.A. El-Safty, F. Mizukami, T. Hanaoka, General and simple approach for control cage and cylindrical mesopores, and thermal/hydrothermal stable frameworks. J. Phys. Chem. B **109**, 9255–9264 (2005)
71. S.A. El-Safty, T. Hanaoka, F. Mizukami, Design of highly stable, ordered cage mesostructured monoliths with controllable pore geometries and sizes. Chem. Mater. **17**, 3137–3145 (2005)

72. T. Kimura, K. Kato, Mesostructural control of non-silica-based hybrid mesoporous film composed of aluminium ethylenediphosphonate using triblock copolymer and their TEM observation. New J. Chem. **31**, 1488–1492 (2007)

73. T.Y. Ma, H. Li, A.N. Tang, Z.Y. Yuan, Ordered, mesoporous metal phosphonate materials with microporous crystalline walls for selective separation techniques. Small **7**, 1827–1837 (2011)

74. T.Y. Ma, Y.S. Wei, T.Z. Ren, L. Liu, Q. Guo, Z.Y. Yuan, Hexagonal mesoporous titanium tetrasulfonates with large conjugated hybrid framework for photoelectric conversion. ACS Appl. Mater. Interf. **2**, 3563–3571 (2010)

75. T. Kimura, Y. Yamauchi, Electron microscopic study on aerosol-assisted synthesis of aluminum organophosphonates using flexible colloidal PS-*b*-PEO templates. Langmuir **28**, 12901–12908 (2012)

76. L.G. Qiu, T. Xu, Z.Q. Li, W. Wang, Y. Wu, X. Jiang, X.Y. Tian, L.D. Zhang, Hierarchically micro- and mesoporous metal-organic frameworks with tunable porosity. Angew. Chem. Int. Ed. **47**, 9629–9633 (2008)

77. T. Kimura, N. Suzuki, P. Gupta, Y. Yamauchi, Effective mesopore tuning using aromatic compounds in the aerosol-sssisted system of aluminum organophosphonate spherical particles. Dalton Trans. **39**, 5139–5144 (2010)

78. S.K. Jana, A. Mochizuki, S. Namba, Progress in pore-size control of mesoporous MCM-41 molecular sieve using surfactant having different alkyl chain lengths and various organic auxiliary chemicals. Catal. Surv. Asia **8**, 1–13 (2004)

79. X.Z. Lin, Z.Y. Yuan, Synthesis of amorphous porous zirconium phosphonate materials: tuneable from micropore to mesopore sizes. RSC Adv. **4**, 32443–32450 (2014)

80. H. Shibata, T. Ogura, T. Mukai, T. Ohkubo, H. Sakai, M. Abe, Direct synthesis of mesoporous titania particles having a crystalline wall. J. Am. Chem. Soc. **127**, 16396–16397 (2005)

81. S. Inagaki, S. Guan, T. Ohsuna, O. Terasaki, An ordered mesoporous organosilica hybrid material with a crystal-like wall structure. Nature **416**, 304–307 (2002)

82. T. Kimura, D. Nakashima, N. Miyamoto, Lamellar mesostructured aluminum organophosphonate with unique crystalline framework. Chem. Lett. **38**, 916–917 (2009)

83. H.Y. Chen, H.X. Xi, X.Y. Cai, Y. Qian, Experimental and molecular simulation studies of a ZSM-5-MCM-41 micro-mesoporous molecular sieve. Micropor. Mesopor. Mater. **118**, 396–402 (2009)

84. Y.J. Wu, X.Q. Ren, J. Wang, Effect of microwave-assisted aging on the static hydrothermal synthesis of zeolite MCM-22. Micropor. Mesopor. Mater. **116**, 386–393 (2008)

85. W.C. Conner, G. Tompsett, K.H. Lee, K.S. Yngvesson, Microwave synthesis of zeolites: 1. Reactor engineering. J. Phys. Chem. B **108**, 13913–13920 (2004)

86. Y.J. Wu, X.Q. Ren, J. Wang, Effect of microwave-assisted aging on the static hydrothermal synthesis of zeolite MCM-22. Micropor. Mesopor. Mater. **116**, 386–393 (2008)

87. M.H. Pham, G.T. Vuong, F.G. Fontaine, T.O. Do, A route to bimodal micro-mesoporous metal-organic frameworks nanocrystals. Cryst. Growth Des. **12**, 1008–1013 (2012)

88. T.Y. Ma, H. Li, Q.F. Deng, L. Liu, T.Z. Ren, Z.Y. Yuan, Ordered mesoporous metal-organic frameworks consisting of metal disulfonates. Chem. Mater. **24**, 2253–2255 (2012)

Chapter 4
Morphological Design of Mesoporous Hybrid Materials

Abstract The well-structured and controllable micro- or macromorphology of porous nanomaterials is of great importance for practical applications. The possibility to fabricate films, spheres, monoliths, and so on has been explored. Different morphologies show distinct potentials in various areas. Spheres can be used in biosensing and chromatograph packing; films are applicable in the areas of catalysis and separation; and monoliths can find fit in photonic devices. Controllable synthesis on both the mesoscale (mesostructure) and macroscale (morphology) is therefore necessary. The assembly of structures and the control of morphologies for mesoporous materials is a concerted campaign and affect each other. The factors that determine the ultimate morphologies of the mesoporous materials include several elements: hydrolyzation and condensation of inorganic precursor species, types of the surfactant molecules, interaction between precursors or their related species and surfactants, additives, and physical techniques. As to mesoporous siliceous materials, one can achieve mesoporous structures with various morphologies such as fibers, films, monoliths, spheres, vesicles, and "single crystals," through manipulating these factors. Notwithstanding, the morphologies of mesoporous nonsiliceous materials are still restricted in several relatively simple types due to the considerably complicated reaction systems and unpredictable coordination between inorganic and organic moieties. Some classical morphologies of mesoporous hybrid materials are thus presented as follows.

Keywords Morphology control · Mesoporous spheres · Mesoporous fibers · Mesoporous films · Mesostructured nanorods · MOF nanoplates · Mesoporous monoliths

4.1 Spheres and Fibers

The sphere is one of the most common shapes for materials, which has minimal volume and maximal surface area. Mesoporous spheres can be directly used in chromatographic substrates, catalysts and carriers, drug delivery, and electrode materials,

© The Author(s) 2015   61
Y.-P. Zhu and Z.-Y. Yuan, *Mesoporous Organic-Inorganic Non-Siliceous Hybrid Materials*, SpringerBriefs in Molecular Science,
DOI 10.1007/978-3-662-45634-7_4

thus drawing wide attention. A modified Stöber method combined with surfactant self-assembly is usually adopted to prepare mesoporous silica microspheres. Instead of employing typical emulsion-templating methods with nonpolar solvents such as trimethylbenzene, a self-templating emulsion route was developed for the formation of a spherically shaped hollow manganese phosphonate-based hybrid with hierarchically porous shells [1]. Drops of liquid organophosphonic claw molecules added to a manganese chloride solution did not disappear immediately but, in contrast, the drops broke up into many smaller nanosized spheres under mild stirring, and an emulsion-like solution was thus generated. The phosphonic droplets became covered with the subsequently growing layers of the manganese phosphonates, which could preserve the initial shape of the nanodrops. Upon heating under hydrothermal conditions, a portion of phosphonic droplets gradually defused out through the formed phosphonate shell to react with the remaining inorganic metal precursors in the mixed solution, and this process was similar to the Kirkendall effect [2]. The continuous supply of phosphonic acid and metal ions exerted a thermodynamic control over the condensation between the inorganic units and the organic moieties. Correspondingly, hollow manganese phosphonate microspheres possessing mesocellular foam structures close to the internal shell and secondary smaller mesostructured pores approaching the surface layers were formed. The progress that resulted in the formation of phosphonate hybrid microspheres is reminiscent of the interfacial emulsion polymerization technique that has been developed for the nanoscaled silver hollow sphere (using n-dodecane/water emulsion) and silica (using an oil/water emulsion) hollow nanosphere synthesis [3, 4]. The average diameter of these hybrid microspheres is approximately 0.5–2 mm (Fig. 4.1). The shell thickness is about 150 nm, which results from the aggregation of nanospherical particles with diameters of 5–35 nm. A novel mesocellular foam structure, akin to mesostructured cellular foam (MCF), can be seen. Secondary mesostructured pores of several nanometers emerge near the inferior pore surface layers.

More recently, on the basis of the water-soluble but ethanol-insoluble properties of DTPMP, we developed a template-free strategy to synthesis organic–inorganic hybrid of cobalt phosphonate hollow nanostructured spheres, exhibiting high efficiency in oxidizing degradation of organic contaminants in the presence of peroxymonosulphate. A facile phosphate-mediated self-assembly methodology has been carried out to prepare mesoporous nickel phosphate/phosphonate hybrid microspheres, showing surface area of $267 \text{ m}^2 \text{ g}^{-1}$ and total pore volume of $0.191 \text{ cm}^3 \text{ g}^{-1}$ [5]. It could be considered that drops of liquid organophosphonic claw molecules added to a manganese chloride solution did not disappear immediately but, in contrast, the drops broke up into many smaller nanosized spheres under mild stirring, and an emulsion-like solution was thus generated. The phosphonic droplets became covered with the subsequently growing layers of the manganese phosphonates, which could preserve the initial shape of the nanodrops. Upon heating under hydrothermal conditions, a portion of phosphonic droplets gradually defused out through the formed phosphonate shell to react with the remaining inorganic metal precursors in the mixed solution. The continuous supply of phosphonic acid and metal ions exerted a thermodynamic control over the condensation between the inorganic units and the organic moieties. Correspondingly, hollow manganese phosphonate

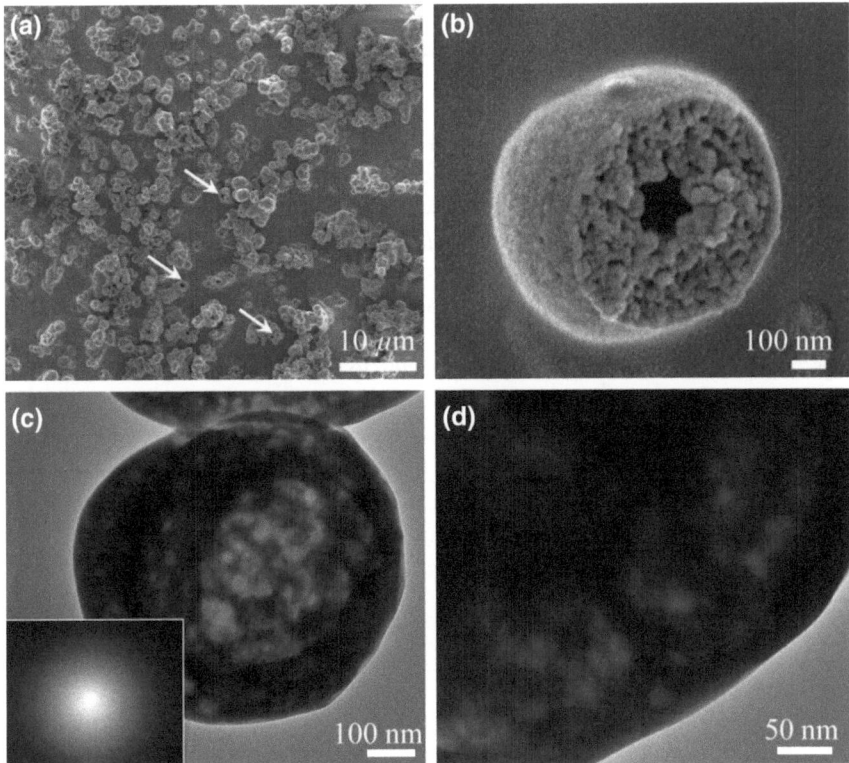

Fig. 4.1 a, b SEM and **c, d** TEM images of hollow manganese phosphonate microspheres with hierarchical porosity. Hollow spheres emphasized with *arrows* in (**a**). Reprinted with permission from Ref. [1]. Copyright 2014, Royal Society of Chemistry

microspheres possessing mesocellular foam structures close to the internal shell and secondary smaller mesostructured pores approaching the surface layers were formed.

Surfactant-templated periodic mesoporous titanium phosphonate materials with alkyleneamine-bridged hybrid frameworks have been synthesized by using the corresponding tetra-phosphonic acid EDTMP as the coupling molecule in an ethanol–water system in the presence of the surfactant Brij 56. By precisely adjusting the composition of the solutions, the spherical morphology with an average diameter of 400–500 nm could be only obtained with the water–ethanol ratio of 3:1 in the reaction system [6]. The resulting materials exhibited a hexagonal (*p6mm*) mesophase, accompanied by a BJH pore width of about 2.2 nm and a specific surface area of 606 m^2 g^{-1}. In addition, irregular macrovoids could be observed throughout the hybrid microspheres, which would facilitate mass transport through the microspheres. The hydrolysis of TiCl$_4$ in the organophosphonate solution gave a multiple component system of organophosphonate–ethanol–water, and thus microemulsion drops were formed under mild stirring. The interfacial microemulsion polymerization of mesophase titanium phosphonate sols rendered the formation of mesostructured titanium phosphonate spheres with homogeneously attached organophosphonate units. In this process, phase separation might take place in the growing aggregates of phosphonate-based mesophases and water/alcohol domains, leading to the creation of sporadic huge macrovoids in hierarchical porous network.

The aerosol-assisted methodology is an effective approach to obtain spherical particles of silica-based and non-silica-based materials, which is based on the sprayed aerosol particles sacrificing and serving as "spherical templates" during the heat treatment process [7, 8]. Spherical aluminum phosphonate particles with uniform mesopores could be fabricated with the assistance of triblock copolymers by temperature-dependent spray-drying (Fig. 4.2a, b) [9, 10]. Pore diameter was widely controlled from 6 to 21 nm by changing surfactants and adding organic additives. In order to obtain periodic mesostructures inside the spherical morphology, the evaporation rate of the solvents (ethanol and water) should be moderate. This could allow the residual soluble species to infiltrate the surfactant scaffolds for the construction of resultant hybrid frameworks with sufficient density [9]. With a further increase in the amount of surfactant molecules with expanded cores (PS-b-PEO), aluminum phosphonates of a fibrous morphology (Fig. 4.2c, d) were

Fig. 4.2 Porous spherical (**a, b**) and fibrous (**c, d**) aluminum phosphonate particles prepared through a spray-drying method in the presence of P123 and colloidal PS-b-PEO templates, respectively. **a, b** Reproduced from Ref. [9] by permission of The Wiley-VCH. **c, d** Reproduced from Ref. [10] by permission of The American Chemical Society

mixed with the spherical ones at a high spray-drying temperature (230 °C) due to the high viscosity of the cloudy precursor solutions [10]. Similarly, Marquez et al. firstly employed the aerosol route to obtain Benchmarked micro- or nanosized HKUST-1, ZIF-8, and $Fe_3(BTC)_2$ (BTC = 1,3,5-benzene tricarboxylate) as well as template-assisted $Fe_3(BTC)_2$ MOFs of various morphologies, and the resulting high space time yields make this continuous method very promising for the industrial production and shaping of MOFs [11].

4.2 Nanoplates and Films

Surfactant molecules can selectively adsorb onto the crystallographic planes to dominate the crystallite growth other than performing as onefold templating agents. In the mixed system of surfactant N-ethyl perfluorooctylsulfonamide ($C_2H_5NHSO_2C_8F_{17}$, N-EtFOSA)–IL 1,1,3,3-tetramethylguanidine (TMGT) for the synthesis of HKUST-1, the Cu^{2+} metal ions dissolved in TMGT react with the deprotonated BTC^{3-} to structure the nanosized framework building blocks [12]. On the one hand, the surfactant molecules play the role of a template in the mesopore formation. On the other hand, the surfactant can selectively adsorb onto the crystallographic planes of the MOF, thus serving as a directing agent and kinetically controlling the anisotropic growth of the MOF. Then the nanosized building blocks with the surfactant N-EtFOSA molecules assemble to form MOF crystals, and the morphologies are determined at the surfactant concentrations. The shapes of the mesoporous MOFs with microporous crystalline walls change from hexagonal via a round to square-shaped nanoplates (Fig. 4.3).

Various desired morphologies could be obtained by controlling the crystallite growth on the confined substrates. With using mesoporous silicas as model supports, highly crystalline homogeneous MOF thin films of HKUST-1 and ZIF-8 combining both micro- and mesoporosity can be successfully synthesized through a layer-by-layer (LBL) methodology [13]. The control over the growth process can be easily achieved by varying the growth cycle numbers. The preferential orientation of the film was along (111), which was due to the interactions between metal ions and the OH groups on the silica foam surface.

As to PMO-type materials, morphological variation is possible via the evaporation-induced self-assembly (EISA) route [14]. On the basis of a solgel process, techniques including spin coating, dip coating, spray coating, and slip casting can be used to prepare the thin films. For spin coating, high-speed rotation-induced centrifugal forces are employed to evenly disperse precursor solution, containing inorganic species and surfactants, on substrates to form thin films. Dip coating is similar to spin coating, but it relies on controlled dipping of substrates in a precursor solution to form the films. These two techniques are most popularly utilized. As to spray coating, an aerosol dispenser or sprayer is used to disperse the precursor solution onto substrates. Slip casting features the substrates containing some pores, whose sizes

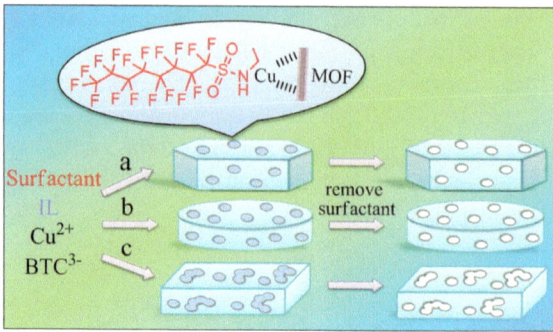

Fig. 4.3 Schematic illustration for the formation of mesoporous MOF nanoplates in surfactant–IL solutions at low (**a**), medium (**b**), and high (**c**) surfactant concentrations. Reprinted with permission from Ref. [12]. Copyright 2012, Royal Society of Chemistry

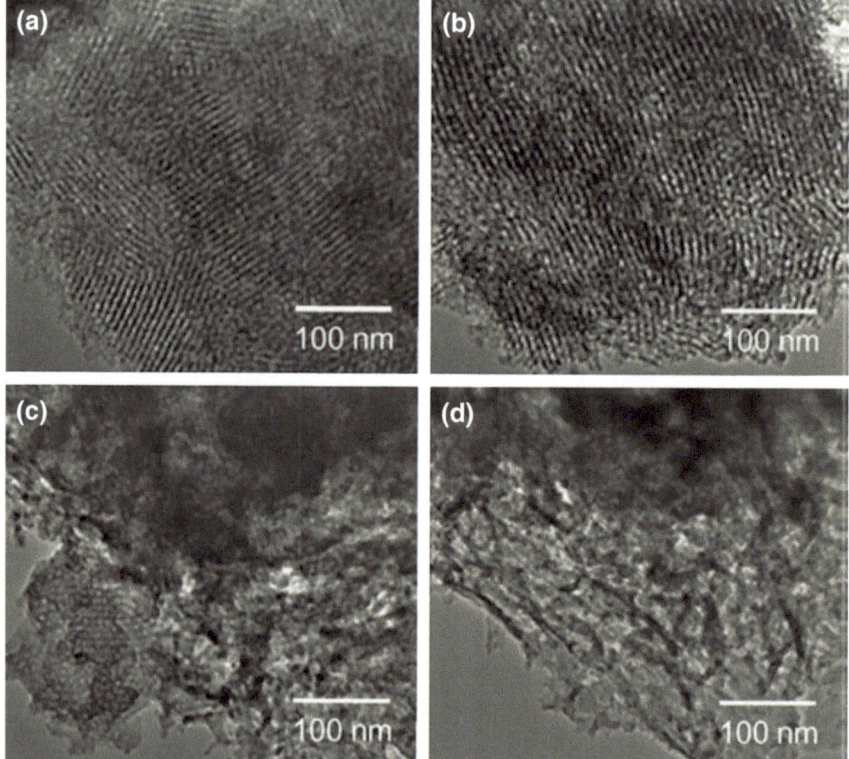

Fig. 4.4 Representative TEM images of 250 °C-calcined mesoporous aluminum phosphonate films prepared using **a** 2.0 g, **b** 2.4 g, **c** 2.8 g, and **d** 3.2 g of $EO_{80}PO_{30}EO_{80}$. Reprinted with permission from Ref. [15]. Copyright 2007, Royal Society of Chemistry

can affect the thickness of thin films. Besides the two approaches, mesoporous thin films can also be made by pulsed laser deposition and electrochemical deposition. In accordance with these strategies, a great diversity of hybrid organic–inorganic materials with controllable morphologies can be achieved by combination with existing techniques. Ordered mesoporous aluminum phosphonate films with high transparency were prepared through the spin coating of an ethanol–water solution containing methylene diphosphonic acid, aluminum chloride, and an $EO_nPO_mEO_n$-type triblock copolymer [15]. The amount of $EO_{80}PO_{30}EO_{80}$ in the precursor solution strongly influenced the mesostructural ordering of the aluminum phosphonate films (Fig. 4.4), and careful heating was essential to maintain the mesostructure after surfactant removal. The mesostructural ordering was gradually decreased with an increase in the added amount of $EO_{80}PO_{30}EO_{80}$, leading to the formation of other phases that are not defined by assemblies of $EO_{80}PO_{30}EO_{80}$. Much larger pores were formed over the entire films and would be formed through a phase separation by the presence of excess $EO_{80}PO_{30}EO_{80}$.

4.3 Nanorods

The formation of phosphonate-based hybrids in a microemulsion system could also result in mesoporous materials with various morphologies. If HEDP and tetrabutyl titanate were used as precursors, the multicomponent microemulsion drops were mainly composed of alkoxide, water, ethanol from the solvent, and butanol from the hydrolysis of tetrabutyl titanate. The interfacial microemulsion polymerization of titanium phosphonate sols and titanium oxo clusters caused the formation of mesostructured titania phosphonate nanorods with a length of 80–150 nm and a thickness of 18–38 nm, possessing homogeneously attached organophosphonate units. The nanorods formed aggregates with the microemulsions to give a hierarchical macroporous structure [16]. The multipoint BET surface area was 257 $m^2\,g^{-1}$, with a BJH pore size of 2.0 nm and a total pore volume of 0.263 $cm^3\,g^{-1}$.

A low-temperature hydrothermal method was developed for the synthesis of lanthanide phenylphosphonate nanorods that are expected to exhibit some novel properties [17]. The mechanism of synthesis and shape control of lanthanide phenylphosphonates is proposed from a kinetic perspective. On one hand, the used p-toluenesulfonate could form loose complex through the electrostatic action with La^{3+} cations to direct the crystal growth process. On the other hand, the addition of p-toluenesulfonate can significantly decrease the viscosity of the solution, which increases the mobility of the components in the system and allows atoms, ions, or molecules to adopt appropriate positions in developing crystal lattices [18].

A coordination modulation method, in which acetic acid is used to directly influence the coordination equilibria, can be used to control the crystal growth of

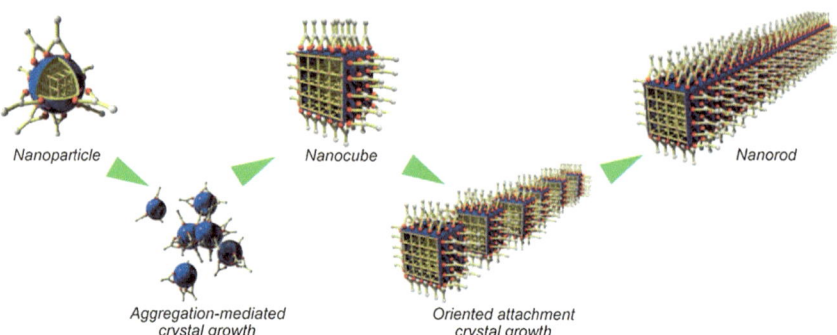

Fig. 4.5 Proposed growth mechanism for $[\{Cu_2(ndc)_2(dabco)\}_n]$ nanorods. The growth process of nanocubes is a consequence of nanoparticle aggregation-mediated crystal growth. The selective coordination modulation on the (100) surfaces of the nanocubes induces the oriented attachment leading the growth of nanorods in the [001] direction. Reprinted with permission from Ref. [19]. Copyright 2009, Wiley-VCH

nanosized $[\{Cu_2(ndc)_2(dabco)\}_n]$ crystals (ndc $=$ 1,4-naphthalene dicarboxylate, dabco $=$ 1,4-diazabicyclo[2,2,2]octane) [19]. The competitive interaction between the coordination mode used to construct the framework and the acetate–copper interaction played a crucial role in determining the reaction rate and crystal morphology. The mechanism of anisotropic crystal growth could be related with the oriented attachment (Fig. 4.5). The coordination modulation method affords perfect framework regularity in the nanocrystals, which allows the nanocrystals to be applied as crystalline porous materials.

Karimi and Morsali successfully used different concentrations of mesoporous silica (SBA-15) for directed growth of the archetypal MOF-5 microcrystals along preferred crystallographic orientations, leading to the formation of interesting micromorphologies [20]. At concentrations as low as 1 wt%, flower-like frameworks are produced, instead of the conventional cubic microcrystals, and unusual nanorods of MOF-5 appear at 3 and 5 wt% of mesoporous silica. Formation of nucleation seeds via linkages between the silanol groups of mesoporous silica and the metallic centers of the MOF is hypothesized to be responsible for oriented arrangement of MOF species. Interestingly, no substrate prefunctionalization is needed to make the SBA-15 a proper structure directing agent, and silanol interactions are found to be sufficient for suitable framework connectivity.

Although non-siliceous nanorods have been prepared through different strategies, the simultaneous introduction of well-structured mesoporosity has been scarcely reported. Different from the conventional inorganic crystalline materials, the unpredictable growth process of non-siliceous materials and the complexity of synthesis systems make them difficult to produce specific micromorphologies with sufficient reproducibility. Correspondingly, new techniques and supermolecules are expected to mediate the nucleation process and thus direct the ultimate morphologies.

4.4 Monoliths

Rigid monoliths are of great significance to optical devices. The molding of the powders into monoliths is also necessary with several advantages including mechanical stability, ease of handling and recovery, and greater structural uniformity to meet the broad needs of industry and household. They can be directly used across the areas of catalysis, sorption, separation, and water treatment.

Periodic mesoporous titanium phosphonate (PMTP-2) monoliths were synthesized by combining autoclaving process and EISA strategy. After experiencing low-temperature hydrothermal aging of the reaction mixture of HEDP and TiCl$_4$ in the presence of oligomer surfactant Brij 56, the complete condensation and coordination of titanium and phosphonic acid could result in the generation of a transparent liquid with great viscosity. The solvent was subsequently evaporated at 50 °C, similar to the EISA method, resulting in titanium phosphonate monoliths (Fig. 4.6) [21, 22]. This might be caused by the polymerization of the organic bridged groups in the network with the inorganic species, making them more like some kinds of macromolecular polymer with mechanical strength and ductility. A mild ethanol exaction process could efficiently remove the involved soft template molecules, leaving highly ordered hexagonal mesostructures [21]. The as-synthesized samples could be molded into various macroscopic morphologies, and valuably, the monolithic shape could be well preserved even after surfactant removal, which may potentially fulfill the qualifications for some industrial devices. The resultant surface area and pore volume were determined to be 1,034 m^2 g^{-1} and 0.51 cm^3 g^{-1}, respectively. The organophosphonate groups were homogeneously incorporated in the framework of the periodic mesoporous hybrid solids, presenting a thermal stability up to approximately 450 °C.

A major challenge for the practical use of MOFs is to deliver them in a suitable shape. However, coordination polymers are mostly obtained and characterized in a powdered crystalline state, and thus typically compounding with binders and pelletizing is required causing a reduction of the inner surface area and pore blocking. On the contrary, coordination polymer gels should provide a versatile access to the synthesis of porous solid bodies of any desired shape at relatively low costs. Aerogels combine both high surface area and good surface accessibility due to their bimodal micro- and macroporous structure. Therefore, they can be utilized as catalyst supports or as catalysts themselves. Kaskel and coworkers reported the synthesis of iron–BTC aerogels with high permanent porosity and total pore volumes of up to 5.6 cm^3 g^{-1} by a solgel approach [23]. The resulting purified gel is placed in an autoclave and subsequently the adsorbed ethanol is exchanged with CO$_2$ for a time of about 24 h to produce monolithic aerogels. Depending on the initial concentration of the solutions, aerogels with densities of 14.5, 32.4, 62.1, and 110.5 mg cm^{-3} are obtained from 0.05, 0.1, 0.2, and 0.4 M trimesic acid solutions, respectively. Figure 4.7 showed an aerogel and the corresponding xerogel that was dried at 40 °C for 48 h. The picture highlights the shrinkage of air-dried gels due to the syneresis.

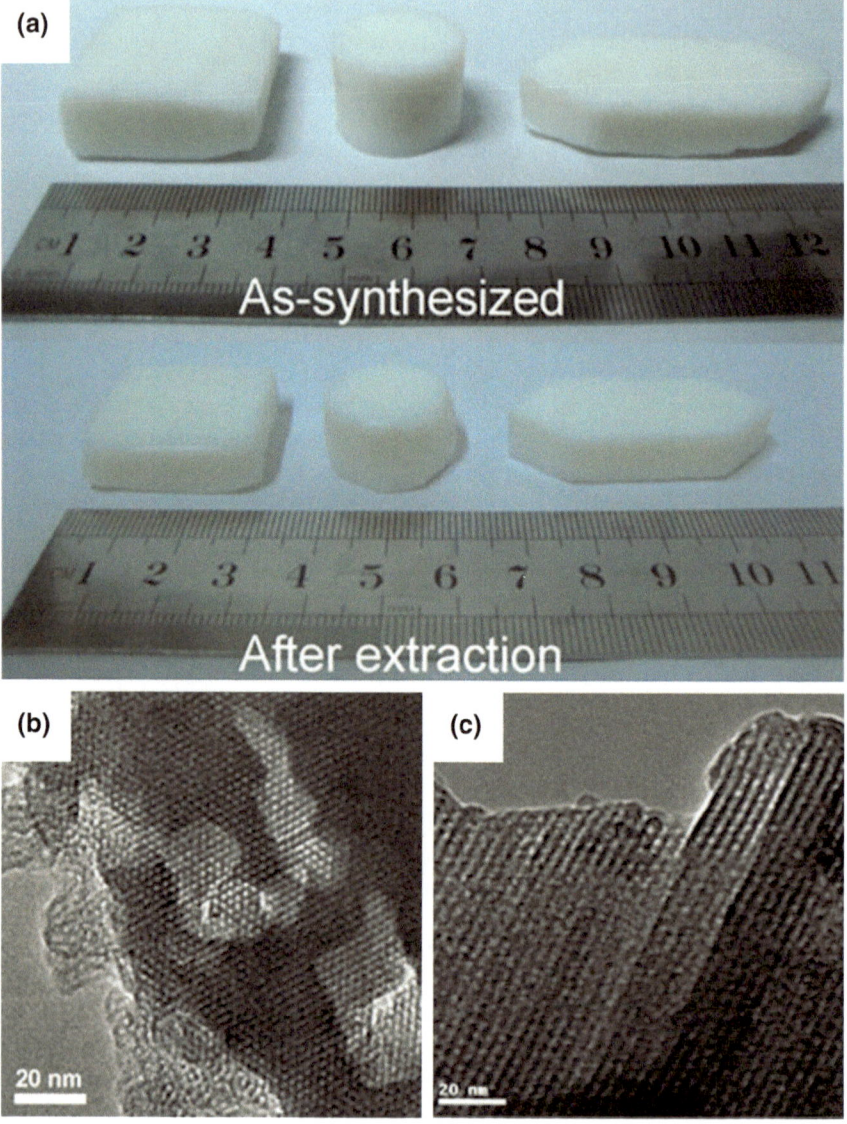

Fig. 4.6 **a** Photographs of as-synthesized periodic mesoporous titanium phosphonate materials and the final monolithic product after surfactant removal by extraction, **b, c** corresponding TEM images. Reprinted with permission from Ref. [21]. Copyright 2010, Royal Society of Chemistry

The main challenges in the preparation of mesoporous monolith materials can be associated with the following points:

Firstly, prevention of cracking. There are two interfaces, gas/liquid (solid) and liquid (solid)/solid (substrate or container) interfaces in the preparation, whose different contraction coefficients often result in cracking of the monolith.

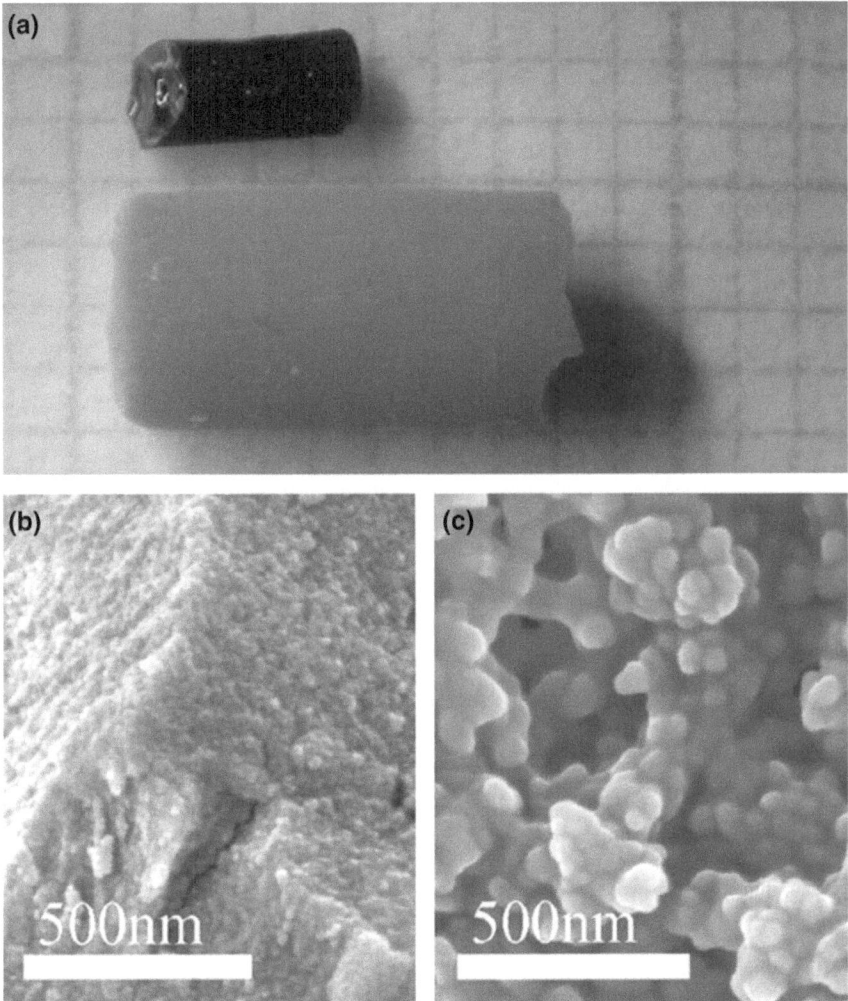

Fig. 4.7 Photograph of an aerogel obtained from 0.2 M trimesic acid solution in comparison with an air-dried xerogel of similar original size (**a**) and scanning electron micrographs (*SEM*) of a xerogel (**b**) and an aerogel sample (**c**). Reprinted with permission from Ref. [23]. Copyright 2009, Royal Society of Chemistry

Secondly, assembly inside monolith bulk phase. When preparing mesoporous films, it is found that surface tension can greatly affect the assembly of mesostructure. Generally, at the gas/liquid (solid) interface of the film surface or the liquid (solid)/solid interface of the bottom, ordered mesostructures are easily organized. While in the middle (inside) of the films, it is highly possible to form disordered mesostructure. For monolith materials, it needs to assemble mesostructures in a bulk middle region, thus the synthetic condition could be more rigid than that of films.

Lastly, maintenance of crack-free morphology after template removal. This point is very important for monoliths, but in most cases, it is very difficult to achieve. Indeed, after removing surfactant templates by extraction or ion exchange, the framework shrinks more or less, inducing the crack of monoliths. Using the CO_2 supercritical method or lyophilization can help to retain integral monoliths to a large extent, but these techniques have been seldom applied in the preparation of mesoporous hybrid materials, further exploration is of considerable value from the scientific point of view.

References

1. Y.P. Zhu, Y.L. Liu, T.Z. Ren, Z.Y. Yuan, Hollow manganese phosphonate microspheres with hierarchical porosity for efficient adsorption and separation. Nanoscale **6**, 6627–6636 (2014)
2. A. Cabot, M. Ibáññez, P. Guardia, A.P. Alivisatos, Reaction regimes on the synthesis of hollow particles by the Kirkendall effect. J. Am. Chem. Soc. **131**, 11326–11328 (2009)
3. C. Kind, R. Popescu, E. Müller, C. Feldmann, Microemulsion-based synthesis of nanoscaled silver hollow spheres and direct comparison with massive particles of similar size. Nanoscale **2**, 2223–2229 (2010)
4. S. Schacht, Q. Huo, I.G. Voigt-Martin, G.D. Stucky, F. Schüth, Oil-water interface templating of mesoporous macroscale structures. Science **273**, 768–771 (1996)
5. Y.P. Zhu, Y.L. Liu, T.Z. Ren, Z.Y. Yuan, Mesoporous nickel phosphate/phosphonate hybrid microspheres with excellent performance for adsorption and catalysis. RSC Adv. **4**, 16018–16021 (2014)
6. T.Y. Ma, Z.Y. Yuan, Periodic mesoporous titanium phosphonate spheres for high dispersion of CuO nanoparticles. Dalton Trans. **39**, 9570–9578 (2010)
7. P.Z. Araujo, V. Luca, P.B. Bozzano, H.L. Bianchi, G.J.A.A. Soler-Illia, M.A. Blesa, Aerosol-assisted production of mesoporous titania microspheres with enhanced photocatalytic activity: the basis of an improved process. ACS Appl. Mater. Interfaces **2**, 1663–1673 (2010)
8. T. Kimura, K. Kato, Y. Yamauchi, Temperature-controlled and aerosol-assisted synthesis of aluminium organophosphonate spherical particles with uniform mesopores. Chem. Commun. **7**(33), 4938–4940 (2009)
9. T. Kimura, Y. Yamauchi, General information to obtain spherical particles with ordered mesoporous structures. Chem. Asian J. **8**, 160–167 (2013)
10. T. Kimura, Y. Yamauchi, Electron microscopic study on aerosol-assisted synthesis of aluminum organophosphonates using flexible colloidal PS-*b*-PEO templates. Langmuir **28**, 12901–12908 (2012)
11. A.G. Marquez, P. Horcajada, D. Grosso, G. Ferey, C. Serre, C. Sanchez, C. Boissiere, Green scalable aerosol synthesis of porous metal-organic frameworks. Chem. Commun. **49**, 3848–3850 (2013)
12. L. Peng, J. Zhang, J. Li, B. Han, Z. Xue, G. Yang, Surfactant-directed assembly of mesoporous metal-organic framework nanoplates in ionic liquids. Chem. Commun. **48**, 8688–8690 (2012)
13. O. Shekhah, L. Fu, R. Sougrat, Y. Belmabkhout, A.J. Cairns, E.P. Giannelisb, M. Eddaoudi, Successful implementation of the stepwise layer-by-layer growth of MOF thin films on confined surfaces: mesoporous silica foam as a first case study. Chem. Commun. **48**, 11434–11436 (2012)
14. H. Fan, Y. Lu, S.T. Reed, T. Baer, R. Schunk, V. Perez-Luna, G.P. López, C.J. Brinker, Rapid prototyping of patterned functional nanostructures. Nature **405**, 56–60 (2000)

15. T. Kimura, K. Kato, Mesostructural control of non-silica-based hybrid mesoporous film composed of aluminium ethylenediphosphonate using triblock copolymer and their TEM observation. New J. Chem. **31**, 1488–1492 (2007)

16. X.J. Zhang, T.Y. Ma, Z.Y. Yuan, Nanostructured titania-diphosphonate hybrid materials with a porous hierarchy. Eur. J. Inorg. Chem. **2008**(17), 2721–2726 (2008)

17. S.Y. Song, J.F. Ma, J. Yang, M.H. Cao, H.J. Zhang, H.S. Wang, K.Y. Yang, Systematic synthesis and characterization of single-crystal lanthanide phenylphosphonate nanorods. Inorg. Chem. **45**, 1201–1207 (2006)

18. B. Tang, L. Zhuo, J. Ge, J. Niu, Z. Shi, Hydrothermal synthesis of ultralong and single-crystalline $Cd(OH)_2$ nanowires using alkali salts as mineralizers. Inorg. Chem. **44**, 2568–2569 (2005)

19. T. Tsuruoka, S. Furukawa, Y. Takashima, K. Yoshida, S. Isoda, S. Kitagawa, Nanoporous nanorods fabricated by coordination modulation and oriented attachment growth. Angew. Chem. Int. Ed. **48**, 4739–4743 (2009)

20. Z. Karimi, A. Morsali, Modulated formation of metal-organic frameworks by oriented growth over mesoporous silica. J. Mater. Chem. A **1**, 3047–3054 (2013)

21. T.Y. Ma, Z.Y. Yuan, Functionalized periodic mesoporous titanium phosphonate monoliths with large ion exchange capacity. Chem. Commun. **46**, 2325–2327 (2010)

22. T.Y. Ma, X.Z. Lin, Z.Y. Yuan, Periodic mesoporous titanium phosphonate hybrid materials. J. Mater. Chem. **20**, 7406–7415 (2010)

23. M.R. Lohe, M. Rose, S. Kaskel, Metal-organic framework (MOF) aerogels with high micro- and macroporosity. Chem. Commun. **28**(40), 6056–6058 (2009)

Chapter 5
Modification and Potential Applications of Organic–Inorganic Non-Siliceous Hybrid Materials

Abstract Applications of organic–inorganic non-siliceous hybrid materials are emerging. But the developments of mesoporous hybrid materials lag far behind the achievements in syntheses. More strictly, there has been no breakthrough yet in practical applications. However, vital commercial applications are in prospect owing to substantial studies on mesoporous non-siliceous hybrid materials, especially on their intrinsic characteristics and extended modification potential. As compared to general bulk analogues, mesoporous materials possess higher specific surface areas and larger porosities, making them more applicable in many fields. Owing to their extensive porosity, adjustable composition, and controllable structures, mesoporous non-siliceous hybrid materials have been developed as multifunctional materials to display versatile and excellent performances beyond the traditional use as catalysts and adsorbents, even contributing to the developments in the fields ranging from energy storage and conversion to medical diagnosis and therapy. The hybrid frameworks also demonstrate the capability of post-decoration for improved performances and extended potential applications.

Keywords Application exploration · Post-synthetic modification · Photocatalysis · Adsorption and separation · Energy storage and conversion · Biomaterials

The typical applications are summarized in Fig. 5.1.

© The Author(s) 2015
Y.-P. Zhu and Z.-Y. Yuan, *Mesoporous Organic-Inorganic Non-Siliceous Hybrid Materials*, SpringerBriefs in Molecular Science,
DOI 10.1007/978-3-662-45634-7_5

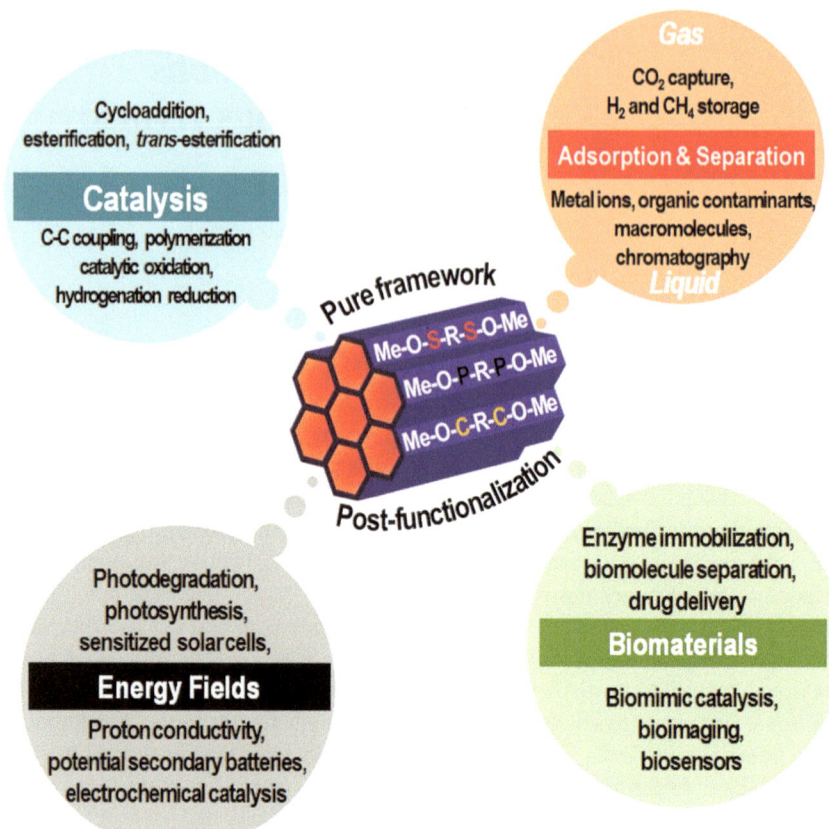

Fig. 5.1 A summary of potential applications of mesoporous non-siliceous hybrid materials

5.1 Multiphase Adsorption and Separation

As one of the most important potentials of mesoporous hybrids, adsorption has attracted much interest. The considerable inner surface area, adjustable pore size, controllable surface compositions, excellent pore accessibility, and the possibility of functionalization make mesoporous non-siliceous hybrid materials ideal for use in the areas of adsorption and separation, for example, in waster water treatment.

5.1.1 Gas Adsorption

As one of the most important potentials of mesoporous hybrids, adsorption has attracted much research interest, such as storage of fuel gases (e.g., hydrogen and methane) and greenhouse gas (e.g., carbon dioxide) capture. Selection of the right molecular building blocks can effectively tune the framework connectivity, pore size, and surface area and thereby optimize its hydrogen sorption capacity. Yaghi and coworkers synthesized a series of mesoporous MOFs with ultrahigh surface area, which exhibited large total hydrogen uptake at 77 K [1]. Zhou et al. prepared mesoporous PCN-105 presenting hydrogen uptake at 1.0 bar of 1.51 wt% at 77 K and 1.06 wt% at 87 K, respectively [2]. Generally, the functionality of the organic linkers has little influence on hydrogen sorption and a large MOF cavity does not make effective contributions to excess hydrogen uptake capacities. Improving the interactions between H_2 and the framework presents a major challenge and bottleneck for MOFs to store hydrogen in a practical manner. An alternative high-density fuel source to hydrogen and gasoline is methane due to its cleaner and more abundant nature. The methane adsorption capacity in micro-/mesoporous UMCM-1 at 298 K could reach 8.0 mmol g^{-1} at 24.2 bar [3]. Noticeably, unlike hydrogen, the interactions between methane and the aromatic hybrid framework are strong enough. The safe, cheap, and convenient means for methane storage are still in its deficiency. Rigorous research toward robust and available mesoporous or even hierarchical porous hybrid materials for large-scale applications is urgently needed.

The capture of greenhouse gases such as CO_2 under practical conditions is quite significant because of the implications for global warming, and the removal of CO_2 from industrial flue gas has become an important issue. One feasible option to curtail the rise of the threats is to capture CO_2 from the combustion of fossil fuels. Among a number of CO_2 capture solids including porous carbons [4, 5], amine-modified mesoporous silicas [6], and carbon–CaO nanocomposites [7], exhibiting certain advantages such as high surface area, large pore volume, uniform pore width, low cost, and relatively high stability is promising for CO_2 capture over a wide range of operating conditions. In recent years, much attention has been focused on mesoporous non-siliceous hybrids for CO_2 capture due to ultrahigh surface area, adjustable surface chemistry, and relatively low cost [8–10]. The CO_2 uptake of the cubic mesoporous titanium phosphonates was approximately 1.0 mmol g^{-1} at 35 °C [11], which was much higher than some pure silica adsorbents and comparable with some amino-modified mesoporous silica with similar surface areas [12]. Theoretically, the incorporation of accessible nitrogen donor groups into the network of porous materials can dramatically influence the gas uptake ability, especially for base carbon oxide [4]. Thus, combined with the superiority of large surface area and high pore volume, the CO_2 uptake capability could be enhanced obviously (about 36.7 wt% at 1 atm and 273 K) when nitrogen-rich organic linkages were intentionally used [13]. Recent theoretical and experimental studies have revealed the correlation between the amount of CO_2 adsorbed and the surface area or pore volume, as well as the increase of adsorption enthalpy for the host materials with open metal sites and active organic functionalities [4, 14]. In particular, the incorporation of mesoporosity and even hierarchical porosity can optimize the adsorption capacity and kinetics. The CO_2 adsorption equilibrium for meso-/macroporous titanium phosphonates (B–Ti–1/2) could be reached within 50 min [15] (Fig. 5.2). The CO_2 uptake was 0.89 mmol g^{-1} at 40 °C, which was comparable with commercially activated carbon [16].

Fig. 5.2 TGA records
of CO_2 adsorption for
the titanium phosphonate
samples, tested at 40 °C.
Reprinted with permission
from Ref. [15]. Copyright
2011, Royal Society of
Chemistry

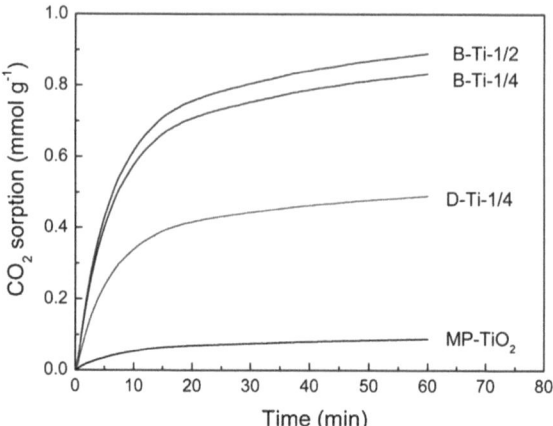

It is noteworthy that the adsorption capacity of porous metal phosphonates remained stable even after multiple cycles, making them promising in practical applications [10]. Increasing the CO_2–framework interactions and improving the porosity of the phosphonate hybrid frameworks present the major challenge and bottleneck in capturing CO_2. Creating open metal sites and introducing basic sites have been shown to effectively increase the interactions. Moreover, in comparison with the carboxylate-based MOFs, the storage of fuel gases such as hydrogen and methane on porous metal phosphonates is scarcely reported. Rigorous investigation toward highly stable and affordable metal phosphonates for theoretical research and practical applications is urgently needed.

5.1.2 Liquid Adsorption and Separation

Adsorption from liquid phase is much more complicated as compared with gas-phase adsorption, due to the competitive behaviors between solute and solvent for the solid surface. The adsorption of a solute is mainly dependent on the molecular sizes and physicochemical properties and on the textual properties and surface chemistry of an adsorbent. It can be envisioned that metal phosphonates can perform as good adsorbents for the removal of heavy metal ions from waste waters, due to that the organic functionalities serve the formation of complexes with metal ions through acid–base interactions, and the easy separation of the loaded solid adsorbent from liquid phase is preferable.

Besides the majority of efforts have been concentrated on the gas adsorption or storage, mesoporous hybrid materials have recently been developed as adsorbents for the removal of heavy metal ions, in which the organic functionalities of these adsorbents serve the formation of complexes with heavy metal ions through acid–base interactions. For instance, the coupling organophosphonic molecules were homogeneously incorporated into the mesoporous walls of the titanium phosphonate materials, and the specific structure of ethylenediamine could chelate metal ions [17]. The adsorption of Cu^{2+}, Pb^{2+}, and Cd^{2+} appeared to follow a Langmuir-type behavior, with the ions being almost quantitatively adsorbed until saturation of the binding sites was reached. The calculated maximum adsorption capacity was 36.49, 29.03, and 26.87 mmol g^{-1} adsorbent for Cu^{2+}, Pb^{2+}, and Cd^{2+}, respectively (Fig. 5.3). The PMTP-1 adsorbent possessed high distribution coefficients (K_d), particularly at low metal ion concentration (67,440 ml g^{-1} when treating 0.2×10^{-4} mol L^{-1} Cu^{2+} solution). The K_d value decreased to 1,517 ml g^{-1} when the Cu^{2+} concentration increased to 1.8×10^{-4} mol L^{-1}, denoting their high affinities for the uptake of Cu^{2+} ions at low levels. Interestingly, the K_d value of Cu^{2+} ions was higher than those of Pb^{2+} (669–52,500 ml g^{-1}) and Cd^{2+} ions (427–50,000 ml g^{-1}), suggesting some selectivity of the PMTP-1 adsorbent to different metal ions. Therefore, the competitive adsorption experiment was also performed by treating a mixed ionic solution of 10 ml containing equal amounts (0.1×10^{-4} mol L^{-1}) of Pb^{2+}, Cu^{2+}, and Cd^{2+} with 20 mg of PMTP-1 sample. The residual concentrations of the metal ions were determined by atomic absorption spectroscopy. The adsorption amounts of 3.21, 2.10, and 2.05 mmol g^{-1} were obtained for Cu^{2+}, Pb^{2+}, and Cd^{2+}, respectively. This proved a distinct preference of the PMTP-1 adsorbent for the uptake of Cu^{2+} ions compared to that of Pb^{2+} and Cd^{2+} ions, indicating that the synthesized periodic mesoporous titanium phosphonate materials have an innate selective affinity for the adsorption of Cu^{2+} over Pb^{2+} and Cd^{2+} ions.

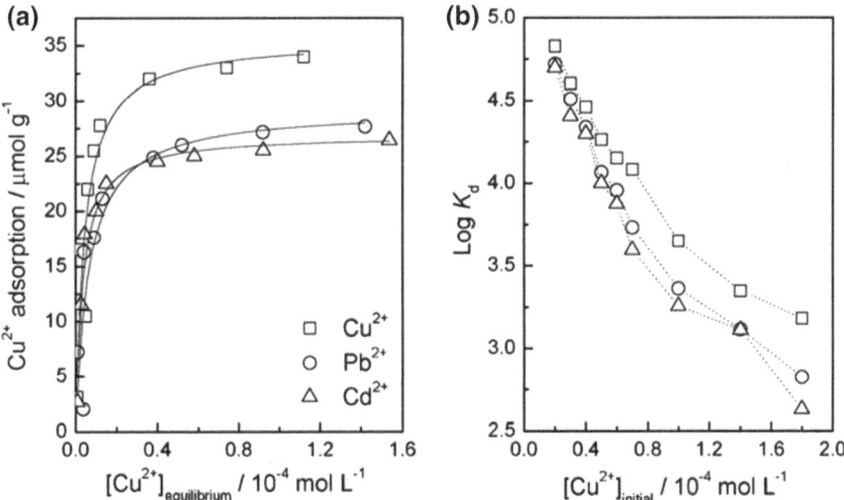

Fig. 5.3 **a** Adsorption isotherms and **b** distribution coefficient profiles of the PMTP-1 adsorbent. The *solid line* in **a** represents a simulation of the adsorption according to the Langmuir equation. Reprinted with permission from Ref. [17]. Copyright 2010, Royal Society of Chemistry

Further experiments confirmed that, besides the ethylenediamine groups, many electronegative groups or atoms including –OH, –SH, and N could contribute to metal ion adsorption. Moreover, competitive adsorption experiments demonstrated that the phosphonate-based adsorbents had an innate selective affinity for the adsorption of one particular ion over the others. A distinct preference of the PMTP-1 adsorbent for the uptake of Cu^{2+} ions, compared with that of Pb^{2+} and Cd^{2+} ions, was observed. This phenomenon was observed in the previous ethylenediamine-containing mesoporous silica, which had also a distinct preference for the uptake of Cu^{2+} ions over Ni^{2+} and Zn^{2+} ions [18]. Therefore, it is important from a technical point of view to select suitable coupling molecules with a specific structure and to enlarge the surface area and pore volume of the hybrid materials, and thus to improve the adsorption performance.

Meanwhile, the adsorption characteristics of organic pollutants on mesoporous metal phosphonate materials were also identified. It was revealed that mesoporous titanium phosphonate PMTP-1 exhibited excellent adsorption performance for the cationic dye methylene blue (MB) as target pollutant from aqueous solution [19]. The adsorption equilibrium was achieved after 30 min of contact time, and the adsorption of MB on PMTP-1 was best fitted to the Langmuir isotherm model with the maximum monolayer adsorption capacity of 617.28 mg g^{-1}, indicating that the PMTP-1 could be used as an efficient adsorbent for the removal of textile dyes from effluents. Results of kinetic studies indicated that the adsorption process followed the pseudo-second-order model, which suggests that the process might involve chemisorption.

The adsorption of biomacromolecules such as proteins from solution onto solid surfaces is also of great scientific importance in many areas, such as biology, medicine, biotechnology, and food processing [20]. Under pH conditions close to the isoelectric point, the adsorption of lysozyme on aluminum phosphonate hybrid materials was dominated by host–guest hydrophobicity–hydrophobicity interactions [21]. Interestingly, unlike inorganic framework adsorbents used for the adsorption of proteins [22], the porous phosphonate hybrid adsorbents had an organic–inorganic framework, which contains plenty of hydrophobic alkyl groups inside the framework [21]. The hydrophobicity/hydrophilicity could be controlled at a chemical dimension. So the hydrophobic interactions between the organic groups inside the channel walls and the nonpolar side chains of the amino acids on the surface of lysozyme were greatly enhanced, leading to an increased monolayer adsorption capacity. When extra-long hydrophobic alkyl chains ($-[CH_2]_6-$) were incorporated, the resultant adsorption ability was higher than for organophosphonates with fewer hydrophobic $-CH_2-$ groups. Since various biomolecules exhibit distinct isoelectric points and spatial sizes, the molecules can be effectively adsorbed and separated by changing the pH and the porosity and pore structures of metal phosphonates.

Chromatography (e.g., gas and liquid phase) is one of the most powerful separative methods in analytical chemistry. PMOs and organically modified mesoporous silicas have been employed as the stationary phases in reverse-phase high-performance liquid chromatography (HPLC) in the form of packed columns, for the separation of only neutral compounds [23, 24]. However, packed columns usually result in low resolution as a result of peak broadening, which impairs the separation efficiency of these hybrid materials. Recently, a series of hexagonal periodic mesoporous metal phosphonates with semicrystalline pore walls synthesized by microwave irradiation were first employed as the stationary phase in the open-tubular capillary electrochromatography (OTCEC) separation technique [25], which combines the efficiency of capillary electrophoresis and the selectivity of HPLC. The presence of functional groups inside the framework and the excellent properties of the obtained metal phosphonates, including well-ordered pores with large surface area and pore volume and high thermal stability, improved selectivity and so encouraged their use as the stationary phase in wall-coated OTCEC separation for various substances including acidic, basic, and neutral compounds. In the case of neutral compounds including benzene, nitrobenzene, naphthalene, and anthracene (Fig. 5.4), the elution order was benzene < nitrobenzene < naphthalene < anthracene on the mesoporous titanium phosphonate-coated capillary, suggesting a hydrophobicity mechanism for the separation of the species. In contrast, for the ordered mesoporous titanium phosphonate constructed from the same phosphonate linkers (EDTMP) but with an amorphous framework, benzene, naphthalene, and anthracene could be well separated, while benzene and nitrobenzene could not be completely separated. This indicated that, in addition to hydrophobic interactions, suitable polarization of the crystalline mesoporous phosphonates is responsible for the effective separation of benzene and nitrobenzene. Since the separation efficiency mainly depends on the host–guest interactions between stationary and mobile phases, such as hydrophobicity–hydrophobicity interaction, polarity, and intermolecular forces, phosphonate hybrid materials, constructed by long alkyl chains [21], multidimensional linkages [26, 27], or phosphonic linkers with polar pendant groups [28, 29], can also be used in chromatography techniques, and this is worth exploring. Three non-, micro- and mesoporous Cd-MOF isomers were used to prepare mesoMOF, and the resultant

Fig. 5.4   OTCEC separation of neutral compounds at pH = 3.0. PMTP-1 and EDTMP-Ti stand for amorphous and crystalline-ordered mesoporous titanium phosphonates, respectively. Reprinted with permission from Ref. [25]. Copyright 2011, Wiley-VCH

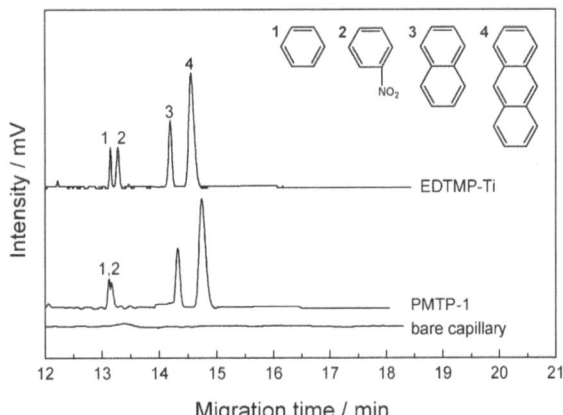

mesoMOFs were employed as stationary phase for liquid chromatography (LC) separations [30]. The open channel of mesoMOF has a dimension of 1.7×2 nm^2, which could facilitate the incorporation of Rh6G (1.3×1.6 nm^2) but exhibit size exclusion of the larger dye Brilliant Blue R-250 (1.8×2.2 nm^2). And the micro-MOF isomer (0.8×1.5 nm^2) with smaller pores could not work.

Mesoporous non-siliceous hybrids have been successfully applied to adsorbing gas, toxic metal ions and organic contaminants, and separating organic compounds, though some further studies are still needed to illustrate the host–guest interactions, such as model fitting and calculation analysis, supplying theoretical basis to direct the synthesis and to optimize the adsorption/separation capabilities. Moreover, the capacity of gas or liquid adsorption/separation is directly related to the surface area, pore volume, pore sizes, and framework composition. Thermal, chemical, and mechanical stabilities are crucial to the practical applications of mesoporous hybrids. Adjusting pore width through using organic groups with demanded length or proper surfactant molecules can make the adsorption or separation of guests with different dimensions feasible. Engineering pore walls with exquisite functionalities enables further potential of mesoporous hybrid materials in separating some special mixtures that cannot be finished by classical porous inorganic solids, for example, liquid/liquid separation and enantioselective separation.

5.2 Energy Conversion and Storage

With the rapid development of human civilization, energy issues have received increasing attention. Coal, petrol, and natural gas as traditional fossil fuels are exceedingly depleted, accompanied with the emission of harmful chemicals to the atmosphere. In response to the energy crisis and environmental contaminations, clean energy and sustainable developments are the basic principles, and renewable solar energy is regarded as an alternative to conventional fossil fuels. Furthermore, there has been significant interest in the development of alternative electronic devices.

5.2.1 Photocatalysis

By introducing specific functional groups into the hybrid materials, photochemical and photoelectric energy conversion can be realized. The photocatalytic process is representative of the utilization of solar energy. Pristine titania is one of the most investigated photocatalysts for environmental remediation and energy storage due to the industrial availability, stability, and appropriate band gap. However, pure titania allows only absorption of the UV portion of the solar radiation. As many guest ions could be introduced into the titania framework via phosphonic modification, mesoporous titanium phosphonates are considered to be potential photocatalysts. In general, the homogeneous doping (C, P, N, etc.) of titanium phosphonates network from the bridging molecules and well-structured porosity could increase the photoadsorption efficiency and enhance the mass transfer. Correspondingly, a noticeable shift of adsorption edge to visible light region was achieved for titanium phosphonate, in contrast to that of pure titania [31]. Interestingly, the band gap energies could be descended by lowering the pH value of the synthesis system [32]. The shift of the absorption onset toward the lower energy range revealed that the titanium phosphonate materials might make a better use of solar energy. The photocatalytic activity of these materials was enhanced because the absorption edge shifting to higher wavelengths, except for the sample synthesized at pH $= 10.5$, which had a small surface area. The extension of the adsorption to visible region of titanium-based phosphonate hybrids resulted in an unprecedentedly impressive photoactivity under the simulated solar light illumination. When the porous structure was adjusted properly with a large surface area of more than $1{,}000\ m^2\ g^{-1}$, the photocatalytic activity could be further enhanced due to the presence of more active adsorption sites [17].

Normally, most of the wastewater produced by industrial process and household contains both organic and inorganic contaminants, such as heavy metal ions and dyes. To cater for the need of the practical applications, a comprehensive catalyst is urgently required [17]. The mixtures of Rhodamine B (RhB), Cu^{2+}, and Pb^{2+} were selected as probes (Fig. 5.5). In the presence of ordered mesoporous titanium phosphonate, the concentration of RhB decreased in the early 2 h and the photodecoloraction rates decreased in the next 6 h until the dyes were completely decomposed. A degradation efficiency of 89.2 % was achieved after 100 min of irradiation, which was even higher than the degree of degradation for RhB on its own (68.4 %) after the same interval. This could be explained by a new broad absorption

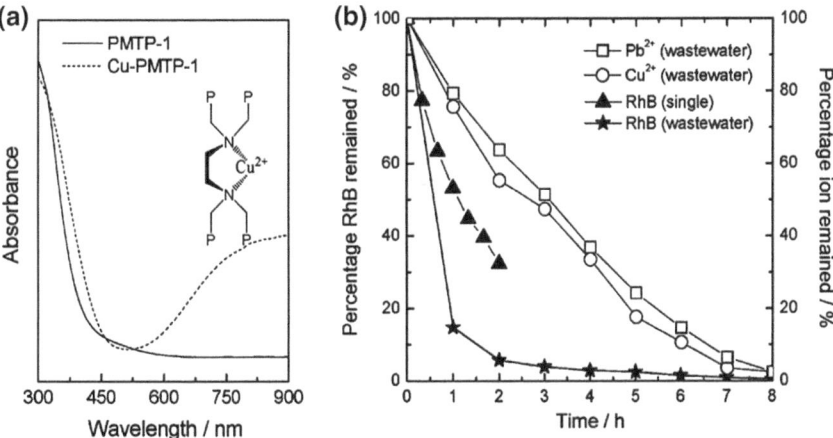

Fig. 5.5 a UV–vis diffuse–reflectance spectra of PMTP-1 before and after Cu^{2+} ion loading; **b** one-pot heavy metal ion adsorption in the wastewater treatment as well as RhB photodegradation using PMTP-1 under simulated solar light irradiation. Reprinted with permission from Ref. [17]. Copyright 2010, Royal Society of Chemistry

peak between 600 and 900 nm caused by the complex of Cu^{2+} on the surface of the hybrid network, giving rise to a better use of visible light. Independently, the concentrations of the metal ions decreased gradually with the reaction time, and nearly, 98 % of ions could be removed after 8-h experiment, which signified that the presence of RhB did not prohibit the adsorption process of the heavy metal ions. Herein, both organic and inorganic pollutants could be eliminated under simulated solar light radiation, while the homogeneously coordinated metal ions on the hybrid materials could improve the photocatalytic activity simultaneously [17]. This inspired us with an alternative method for the preparation of new photocatalysts.

Improving the separation efficiency of photoinduced charge carriers is the basic rule for designing new photocatalytic systems. In general, there are two strategies [33]. Doping with metallic and nonmetallic elements was considered to reduce the band gap for wide band gap metal oxides [34]. The other efficient approach is coupling with other materials so as to build a heterojunction structure at the interface to enhance the separation efficiency of photogenerated electron–hole pairs during the photocatalytic process. Mesoporous phosphonated titania hybrid materials were prepared with the use of amino trimethylene phosphonic acid (ATMP) as the coupling molecule and triblock copolymer F127 as the template [35], in which the phosphonate groups homogeneously anchored on the mesoporous titania, allowing monolayer adsorption of Zn^{2+} by extensive coordination with the organic bridging groups (Fig. 5.6). The highly dispersed photoactive ZnO nanoparticles were then formed through low-temperature annealing (180 °C) of the Zn^{2+} adsorbed mesoporous phosphonated titania, and the resultant ZnO-coupled mesoporous phosphonated titanium oxide photocatalysts exhibited excellent photocatalytic activity and stability in the photodegradation of RhB under both UV and visible light irradiation [36]. In comparison with the pristine mesoporous phosphonate titania, the commercial titania P25, and the ZnO/mesoporous titania prepared by conventional impregnation, the superior photocatalytic performance and stability of the coupled catalyst of ZnO nanoparticles highly dispersed on the mesoporous phosphonated titania might be due to the coupling effect, the well-defined mesoporosity and the incorporation of phosphonic moieties into the TiO_2 network, presenting potential applications in the fields of environmental remediation and solar cells.

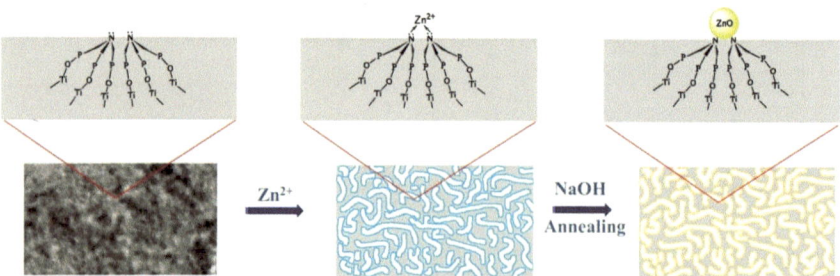

Fig. 5.6 Typical synthesis procedure of the ZnO-phosphonated-TiO_2 mesoporous-coupled photocatalysts. Reprinted with permission from Ref. [36]. Copyright 2014, Elsevier

Particularly, MOFs have demonstrated great potential in artificial photosynthesis since the organization of different molecular components can be realized [37]. The charge-separated excited states of the chromophoric building blocks created upon photon excitation can migrate over long distances to be harvested as redox equivalents at the MOF/liquid interfaces via electron transfer reactions or can directly activate the substrates that have diffused into the MOF channels for photocatalytic reactions [38]. García and coworkers reported the application of the highly stable Zr-containing MOFs (NH₂–UiO–66(Zr)) in photoactive hydrogen production under UV irradiations [39]. By incorporating [Re^I(CO)$_3$Cl(5,5'-dcbpy)] (dcbpy = 2,2'-bipyridine-5,5'-dicarboxylic acid), the active electrocatalyst for CO_2 reduction, into a highly stable Uio-67(Zr) framework, Lin et al. have successfully prepared an active photocatalyst for CO_2 reduction with an obvious higher turnover number (TON) than that of the homogeneous complex [40].

As heterogeneous photocatalysts, MOFs are even superior to inorganic semiconductors in that the effective use of solar light by MOFs can be more facilely achieved by modifications on the metal ions or the organic ligands due to the versatile coordination chemistry of the metal cations, the availability of different organic linkers, and the possibility to modulate the composition, structure, and thus the properties of the MOFs. By taking advantage of the high tunability of the MOF materials, NH₂–Uio–66(Zr) with mixed 2-aminoterephthalic acid (ATA) and 2,5-diaminoterephthalate (DTA) ligands was prepared and was demonstrated to show higher performance for photocatalytic CO_2 reduction due to its enhanced light adsorption and increased adsorption of CO_2 [41]. Rosseinsky and coworkers recently reported the synthesis of a water-stable porous porphyrin MOF (Al-PMOF) by treating AlCl₃ with free-base mesotetra(4-carboxyl-phenyl)porphyrin under hydrothermal conditions [42]. The four carboxylate groups of each porphyrin linker of the Al-PMOF coordinate to eight Al centers, while the metal centers form an Al(OH)O₄ chain bridged by carboxylate oxygen atoms and μ_2-OH⁻ moieties. The connectivity of the Al-PMOF is similar to that of MIL-60 [43]. The Al-PMOF is photoactive with visible light excitation and has been evaluated for the visible-light-driven hydrogen generation from water (Fig. 5.7). When the proton reduction was carried out using the Al-PMOF/MV^{2+}/EDTA/Pt system that

Fig. 5.7 Proposed photocatalytic reactions using Al-PMOF in the presence of MV (*i*) or in the absence of MV (*ii*). Reprinted with permission from Ref. [42]. Copyright 2012, Wiley-VCH

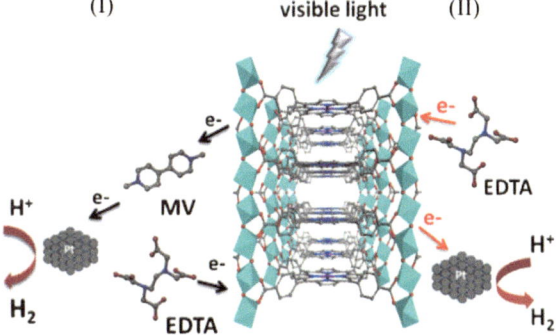

is analogous to the strategy used by Mori et al., low activity was found with much less than stoichiometric amounts of H_2 generated for each porphyrin strut. The authors believed that the low activity resulted from slow diffusion of methyl viologen through the Al-PMOF channels, which leads to ineffective electron transfer from the reduced viologen radical to Pt nanoparticles. Interestingly, when methyl viologen was removed from the reaction mixture, the rate and quantum yield of hydrogen generation increased to about 200 μmol g^{-1} h^{-1} and about 0.1 %, respectively, more than 1 order of magnitude higher than the MV-based approach. In the MV-free approach, about 0.7 H_2 molecules were generated for each porphyrin strut in 6 h of photocatalytic reaction. The Al-PMOF frameworks remained intact after the photocatalytic reaction, as revealed by powder XRD and further confirmed by no detectable leaching of the porphyrin strut into the solution. The ability to post-synthetically metalate the porphyrin strut in this system should allow further tuning of the photocatalytic activity to increase hydrogen generation TONs.

All these studies demonstrate the high potential of MOFs as promising photocatalysts. Photoactive MOFs have shown interesting potential in solar energy utilization, though research on photocatalysis based on MOF materials is still in its infancy. They provide a promising platform to integrate different functional molecular components to achieve light harvesting and photocatalysis. Further studies focused on the optimization of porous structures are worthy of efforts as well. Photoactive non-siliceous hybrids are believed to receive increasing attention from both synthetic chemists and material scientists.

5.2.2 Photoelectrochemical Conversion

Dye-sensitized solar cells (DSSCs) have been widely investigated since 1991 [44]. Up to now, by subtly designing the organic photoactive dye molecules, judiciously choosing electrolytes and optimizing assembling technology, the energy conversion efficiency could be promoted to as high as approximately 15 % [45]. Typical SSCs consist of three parts: working electrodes, counter electrodes, and liquid or polymeric electrolytes (Fig. 5.8). As regards the working electrode, TiO_2 is the most commonly used semiconductor, performing as an electron selective layer between the photosensitizers (organic dyes and quantum dots) and electron-collecting conducting glasses. Important aspects for optimization of the cell performance are selection of the photosensitizers and its attachment motif to the semiconductor surface. Several of the most efficient dye-sensitized SCs (DSSCs) contain ruthenium–polypyridyl complexes as dyes, though the costly dyes and complicated fabrication procedure are typically involved, which prohibit their practical potential. On the other hand, traditional preparation for the dye-sensitized electrodes of solar cells is accomplished by the adsorption of dye molecules onto the presynthesized semiconductor electrodes, which usually leads to a very low loading amount of the photosensitive molecules. An alternative strategy for the construction of new DSSCs was proposed using hybrid metal sulfonate mesoporous materials with large conjugated hybrid

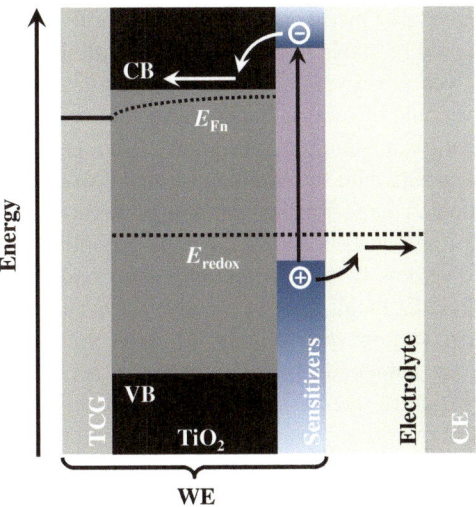

Fig. 5.8 Schematic illustration of sensitized solar cells. A typical working electrode (*WE*) is composed of transparent conductive glass (*TCG*), TiO$_2$, and sensitizers (organic dyes and quantum dots). The sensitized nanostructures are immersed in redox electrolyte, and the circuit is closed by a counter electrode (*CE*). The latter is usually illuminated through a counter electrode (*CE*). Energy band diagram showing the conduction- (*CB*) and valence-band (*VB*) edges of the wide band gap semiconductor (e.g., TiO$_2$), the ground and excited level of the sensitizers and the redox potential Eredox. Upon solar light illumination, electrons are injected from the excited state into the TiO$_2$, while the oxidized QD is recharged by the redox electrolyte

framework (Fig. 5.9) [46]. Ordered hexagonal mesoporous titanium tetrasulfonate materials (CuPcS$_4$–Ti) were synthesized through a hydrothermal process with the assistance of surfactant F127, using the copper(II) phthalocyanine–tetrasulfonic acid tetrasodium salt (CuPcS$_4$) as coupling molecules. It was confirmed that the CuPcS$_4$

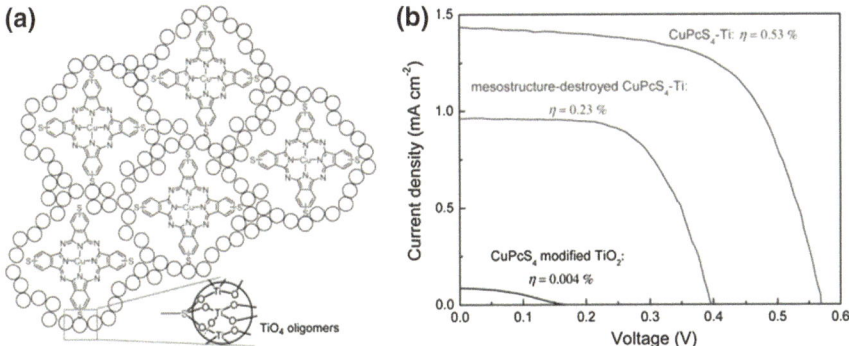

Fig. 5.9 a Proposed skeletal structure of the synthesized CuPcS$_4$–Ti material, **b** current–voltage characteristics of the mesoporous CuPcS$_4$–Ti-based solar cell under simulated sunlight irradiation. Reprinted with permission from Ref. [46]. Copyright 2010, American Chemical Society

groups were homogenously incorporated into the hybrid framework, and the synthesized materials could be stable to around 328 °C with the hybrid framework and ordered mesopores well preserved. Substituted dye molecules like phthalocyanines with sulfonic groups could be used as the coupling molecules. The one-pot condensation between metal precursors and dyes allowed the molecular-level penetration of large π-aromatic groups into the semiconductor network homogenously, resulting in an unprecedented large loading amount of organic dyes, but without the disadvantages of dye aggregation and electron ill-transmission because of the isolation of single dye centers by the surrounding semiconductor oligomers. A high dye content of Ti/CuPcS$_4$ molar ratio at around 50 was achieved, which could be useful in the photoelectric conversion applications. A novel model of isolated dye centers surrounded by semiconductor oligomers was set, which could effectively suppress the aggregation of dye molecules that may decrease the conversion efficiency in some traditional dye-sensitized solar cells. It was proved that the synthesized CuPcS$_4$–Ti exhibited a relatively high conversion efficiency of 0.53 %. More metal phosphonates, sulfonates, and carboxylates with large conjugated structures are still expected. This model supplies us with an alternative strategy for the construction of new dye-sensitized solar cells from organic–inorganic hybrid mesoporous materials.

Though efficient in terms of good interfacial electronic coupling in DSSCs, dyes bearing carboxylate anchors have shown limited stability in aqueous and highly oxidizing conditions [47], and bifunctional long-chain carboxylic acids tend to form undesirable looping structures. Very recently, phosphonic acids were found to offer a promising alternative owing to their high affinity toward the surfaces of metal oxides and the relatively stronger binding than carboxylic acids [48, 49], and they would thereby give better long-term stability of DSSCs. Mulhern et al. [50] have analyzed the influence of the surface-attachment functions of the dyes on electron transfer at the dye–TiO$_2$ interface and long-term stability. Chalcogenorhodamine dyes were attached to the surface of nanocrystalline TiO$_2$ through phosphonic or carboxylic acid functions. No significant changes in the photoelectrochemical performances of DSSCs were observed. H aggregation (i.e., plane-to-plane π-stacking, which broadens the absorbance and causes a blue shift) and electron transfer reactivity were observed when varying the nature of the anchoring group. However, phosphonic linkers were found to enhance the dye–TiO$_2$ bond stability, particularly upon immersion of the material in acidified acetonitrile. Carboxylic-functionalized dyes desorbed completely from TiO$_2$ within 30 min, while no more than 20 % desorption occurred with phosphonic-functionalized dyes after 2 days of immersion under the same conditions. By varying the solvent, pH, electrolyte, semiconductor, and presence of oxygen, Hanson et al. [47] carried out a series of investigations of the relative desorption of –PO$_3$H$_2$ versus –COOH substituted [Ru^{II}(bpy)$_3$]$^{2+}$ under different conditions. Carboxylic-based dyes were found to detach between 5 and 1,000 times faster than their phosphonic counterparts in all tested media.

Nonetheless, in addition to the fact that phosphonic acids present more stable alternatives, the charge injection rates can be prohibited to some extent due

Fig. 5.10 Ruthenium dye functionalized by both phosphonic and carboxylic acids was anchored on TiO$_2$ substrate. Reprinted with permission from Ref. [52]. Copyright 2013, American Chemical Society

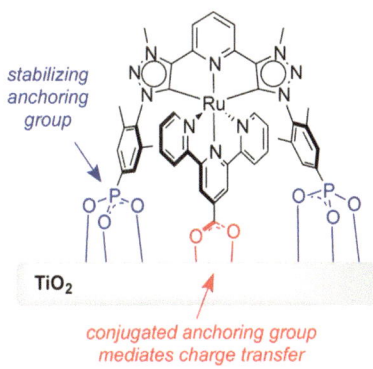

to the tetrahedral phosphorus center and loss of conjugation [50, 51]. Combining the superior binding stability of phosphonate and the good electron injection efficiency of carboxylate has resulted in a feasible method (Fig. 5.10) [52]. A bis(tridentate)-ruthenium complex containing phosphonic and carboxylic acids was elaborated [51]. The underlying basis of this strategy was that the carboxylate moiety only needed to be positioned on the ligand that was involved in the charge transfer to the TiO$_2$, while the phosphonate moieties could be installed on the opposing ligand that did not need to participate directly in the injection process. This led to interesting electron injection properties from the dye into the semiconductor and a good stability in aqueous media. The absolute power conversion efficiencies (PCEs) were not remarkable, which was expected because of the poor spectral coverage of the dyes, but the *trends* in the data provide indirect information about how charge collection is affected by the dye structure.

In the context of solar energy conversion, quantum-dot-sensitized solar cells (QDSSCs) are a promising alternative to existing photovoltaic technologies due to the tunable band gap and promise of stable, low-cost performance [53]. In addition, the QDs open up a way to utilize hot electrons and to generate multiple electron–hole pairs with a single photon through impact ionization. The use of organic linkers between the QDs and titania provides a means of eliminating recombination and leads to an increased conversion efficiency and improved stability [53, 54]. Ardalan et al. [55] investigated the effects of self-assembled monolayers with phosphonic acid head groups on the bonding and the performance of cadmium sulfide (CdS) SSCs. Several organophosphonic acids with different tail groups (–NH$_2$, –COOH and –CH$_3$) were taken as the linkers. It was demonstrated that the nature of the tail group does not significantly affect the uptake of CdS quantum dots on TiO$_2$ nanocrystallites nor their optical properties, but the presence of the phosphonic-based linkers had a significant effect on the photovoltaic device performance. The PCEs in devices made with phosphonic acids were up to about 3 times higher than those without any anchoring agent, which might be due to the organic linkers acting as recombination barriers or the passivated defects at the TiO$_2$ surface (Fig. 5.11) [55]. However, the electrical measurements showed

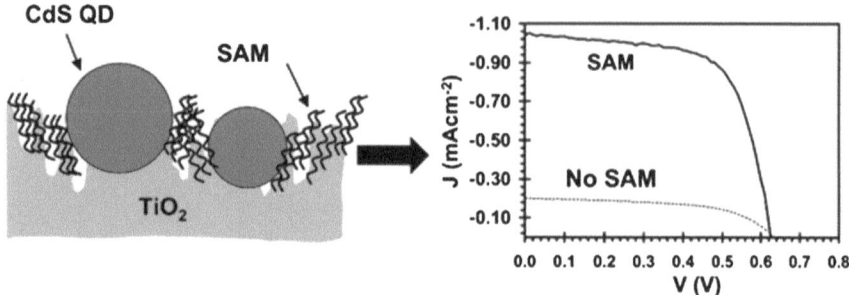

Fig. 5.11 Schematic illustration of the QDs sensitized photoanode in the presence of self-assembled monolayers (*SAMs*) with phosphonic acid head groups on the bonding and the corresponding current–voltage *curves*. Reprinted with permission from Ref. [55]. Copyright 2011, American Chemical Society

that the highest J_{sc} (~1.1 mA cm^{-2}) and power conversion efficiency (~0.44 %) were achieved. Furthermore, the electron injection yield depends on the distance between QDs and TiO$_2$, and it decreases with the increase of linkage chain length [56, 57]. This is a factor worth considering in understanding the functionality of phosphonic linkers and rational design of better photoelectrochemical materials.

Sensitized solar cells play an indispensable role in sustainable development and the exploration of clean energy. It is noteworthy that phosphonate-based DSSCs and QDSSCs show inadequate photolight conversion efficiency, though the corresponding stability of the electrodes shows potential for long-term use. Since QDs can effectively capture solar energy due to the size-dependent absorbance, QD–dye cosensitized solar cells can be worthy of investigation. This can not only make full use of sun light, but also combines the advantages of QDs and organic dyes. A functionalized pore system for energy conversion and storage can be derived from either inorganic components or organic bridging groups with fine photosensitivity, the synergistic effect between which can further improve the ultimate performance. So mesoporous hybrid materials have provided a promising platform for solar energy utilization. If catalytically competent Ir, Ru, and Re complexes with functional organic linkers were introduced into the hybrid framework, the final materials could be used to catalyze water oxidation and CO$_2$ reduction. It can be imagined that a plenty of valuable efforts will be contributed to developing and inventing photosensitive, conductive, or even redox-active porous hybrids for energy conversion and storage in the coming years.

5.2.3 Potential Fuel Cell Applications

Extensive research has been devoted to realizing polymer-supported electrolyte membrane fuel cells consisting of perfluorosulfonic acid polymers (e.g., Nafion);

however, these polymers have some drawbacks, such as operation temperature, humidity, and cost [58]. The use of metal phosphonates as inexpensive proton-conducting membranes for fuel cell applications represents a rising research direction [28, 59]. The phosphonate groups with three oxygen atoms can coordinate with metal ions into multidimensional hybrid frameworks, while the oxygen atoms may still be available to further perform as hydrogen-bonding acceptors [29]. These sites can serve to anchor carrier molecules or directly transfer protons as part of a conduction pathway. Taylor et al. reported the PCMOF3 with a layer structure, $Zn_3(L)(H_2O)_2 \cdot 2H_2O$ (L = [1,3,5-benzenetriphosphonate]$^{6-}$), in which the phosphonate and Zn^{2+} ions did not saturate each others coordination spheres [51]. Thus, the interlayer region was abundant in phosphonate oxygen atoms and Zn-ligated water molecules. The resultant proton conductivity in H_2 was measured as 3.5×10^{-5} S cm^{-1} at 25 °C and 98 % relative humidity (RH). An Arrhenius plot gave a low activation energy of 0.17 eV for proton transfer, indicating the Grotthuss hopping mechanism [60].

Conductivity is a product of the magnitude of the charge, the number of charge carriers, and the mobility of the charges. Conductivity can be tuned by introducing acidic and hydrophilic units, such as carboxylate, phosphonate, and sulfonate groups, due to the presence of hydrophilic oxygen atoms acting as hydrogen-bonding acceptors. After the $C3$-symmetric trisulfonate ligand in PCMOF2 (trisodium 2,4,6-trihydroxy-1,3,5-trisulfonate benzene) was isomorphously substituted with the $C3$-symmetric tris(hydrogen phosphonate) ligand, the resulting material PCMOF21/2 had its proton conduction raised 1.5 orders of magnitude compared to the parent material, to 2.1×10^{-2} S cm^{-1} at 90 % RH and 85 °C, while maintaining the parent MOF structure (Fig. 5.12) [61]. This was due to the pores being partially lined with the hydrogen phosphonate groups rather than exclusively

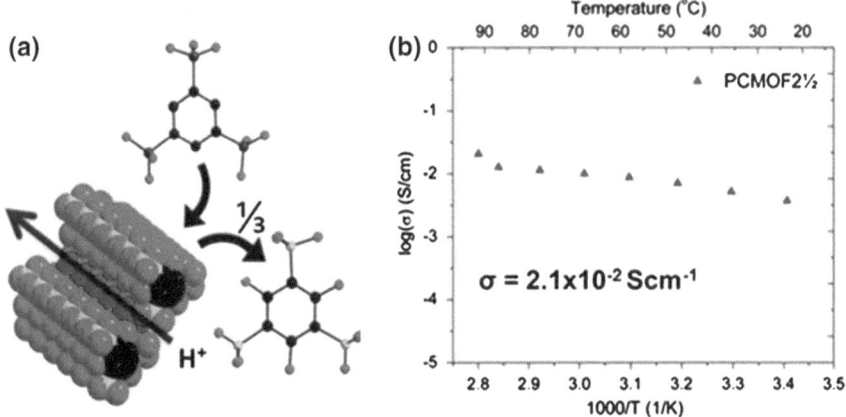

Fig. 5.12 **a** Structure of PCMOF21/2 showing a single pore and space filling cross section of a pore, **b** the corresponding proton conductivity data (90 % RH). Reprinted with permission from Ref. [61]. Copyright 2013, American Chemical Society

non-protonated sulfonate groups, which should augment proton conduction. A series of MOF–polymer composite membranes exhibited an enhanced low-humidity proton conductivity, compared with that of pure MOF submicrometer crystals, $\{[Ca(D\text{-}Hpmpc)(H_2O)_2] \cdot 2HO_{0.5}\}_n$, at 25 °C and about 53 % RH [62]. It was found that the available proton carriers in the MOF structure provided a basis for the conductivity, and the large humidification effect of PVP with adsorbed water molecules greatly contributed to the proton transport in the composite membrane.

Phosphonate-based MOFs have gradually attracted scientists' interest. As discussed above, the efficient preparation of porous crystalline metal phosphonates still presents many difficulties. If metal phosphonates are to serve as proton conductors for practical application, it would be preferable that they function under relatively mild conditions (e.g., at low temperatures and in anhydrous conditions). It is considered that this goal may be achieved through preprotection or post-functionalization of the phosphonic bridging groups.

Notably, the considerable ion-exchange capability of metal phosphonates has been confirmed [63, 64]. Zirconium tetraphosphonates possess an open framework structure with 1D cavities decorated with polar and acidic $P = O$ and P–OH groups [65]. In addition to the excellent proton conductivity, the hybrid was fully protonated by adding HCl and then subjected to several acid–base ion-exchange reactions with alkaline metal ions, such as Li^+, Na^+, and K^+. Anionic MOF of $Zn_{2.5}(H)_{0.4-0.5}(C_6H_3O_9P_3)(H_2O)_{1.9-2}(NH_4)_{0.5-0.6}$ was synthesized with the use of urea and 1,3,5-benzenetriphosphonic acid [66], in which ammonium ions are exchangeable with Li^+. Due to a certain degree of flexibility of the hybrid framework, a reversible insertion/desertion of Li^+ through the pores and elastic network can be envisioned, showing potential for secondary batteries. Although this aspect is not extensively studied, the intrinsic porosity within the conductive hybrid materials (ions or protons) remains largely unknown but worthy of research effort.

Crystalline porous MOFs with hydrated water possess interesting proton-conducting properties, but only at ambient temperatures [67, 68]. The synthesis of coordination polymers with high-temperature proton conductivity can be generally divided into two distinct approaches. First, inherently acidic frameworks can be obtained either by self-assembly of the corresponding functionalized ligands or by post-synthetic modifications of the MOFs. These result in reticular structures with covalently attached acidic groups decorating the pores of the extended coordination network [69, 70]. The alternative way is to imbue the pores of coordination polymers with nonvolatile guest molecules as a medium that provides multiple proton delocalization pathways for efficient proton transport. Generally speaking, the ionic conductivity depends on the amount and mobility of charge carriers (protons). Therefore, the inclusion of stronger acids into porous structures should greatly improve the proton-conducting properties of such hybrid materials. Ponomareva et al. [71] reported the impregnation of the mesoporous MIL-101 by nonvolatile acids H_2SO_4 and H_3PO_4. Such a simple approach afforded solid materials with potent proton-conducting properties at moderate temperatures, which was critically important for the proper function of onboard automobile fuel cells. These hybrid compounds demonstrate high proton conductivity (σ) over a broad temperature range. In fact, the achieved σ values of 1×10^{-2} S cm^{-1} at 150 °C and $\sigma = 3 \times 10^{-3}$ S cm^{-1} under ambient conditions not only beat those of any other MOF-based compounds but are among the highest values reported to date for proton-conducting materials. The confirmed framework and chemical stability at elevated temperatures make such materials promising for automobile fuel cell PEM applications.

Recently, it was showed that porous MOFs could perform as outstanding templates and/or precursors to fabricate porous carbons and related nanostructured functional materials based on their high surface areas, controllable structures, and abundant metal/organic species in their scaffolds [72, 82]. ZIF-7, as another member of ZIFs, was employed as a self-sacrificed precursor with environmentally friendly glucose as an additional carbon source to produce nitrogen-doped porous carbon (Fig. 5.13) [73]. The addition of the environmentally friendly carbon source glucose not only improves the graphitization degree of samples, but also favors removal of Zn metal and zinc compound impurities from ZIF, leading to the formation of metal-free in situ nitrogen-doped porous carbons. Compared to other porous carbons, the Carbon-L (the sample was derived from glucose/ZIF-7 composites that were synthesized under liquid condition) exhibits not only much higher electrocatalytic activity, which is close to that of commercial 20 % Pt/C, but also better stability and increased tolerance to the methanol crossover effects, which is superior to the 20 % Pt/C catalyst. Results indicate that both high electrical conductivity and the content of pyridinic N of the prepared Carbon-L play a key role in electrocatalytic activity for ORR.

Pt containing MOF-253 was synthesized and subsequently subjected to pyrolytic carbonization under non-reactive gas atmosphere [74]. Upon pyrolysis, Pt nanoparticles embedded in electronically conductive carbon media were produced.

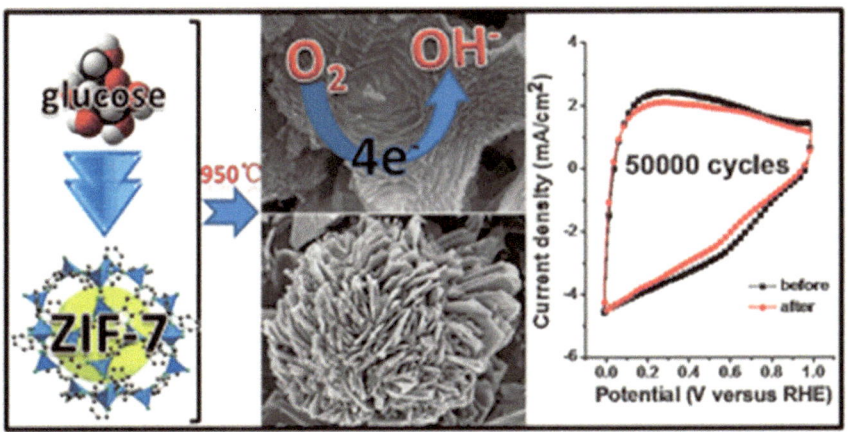

Fig. 5.13 ZIF-7/glucose composite-derived nitrogen-doped porous carbons as metal-free electrocatalysts for ORR exhibiting excellent electrocatalytic activity and operation stability. Reprinted with permission from Ref. [73]. Copyright 2013, Royal Society of Chemistry

The membrane electrode assemblies (MEAs) made of these electrocatalysts were tested as both anode and cathode in a H$_2$/air single cell fuel cell. As the anode, the most promising electrocatalyst (C$_3$) demonstrated an open circuit voltage of 970 mV and power density of 0.58 W mg$_{Pt}^{-1}$ comparable to the commercial electrode power density (0.64 W mg$_{Pt}^{-1}$) at 0.6 V in a single cell test. At the cathode, a power density of 0.38 W mg$_{Pt}^{-1}$ at 0.6 V was achieved. This confirms the promising potential of this simple approach to be used as a technique to prepare efficient fuel cell electrocatalysts.

In fact, a new family of highly porous carbons and composites derived from metal–organic frameworks, known as MOF-derived carbons (MDCs), are attracting tremendous interest for clean energy and environmental applications recently, including hydrogen purification and storage, electrodes for secondary batteries, supercapacitors, electrochemicocatalysis, vapor or gas sensing, carbon capture, and gas separation. Given the flexibility in designing a wide variety of precursor MOF structures with well-structured porosity, functional framework ligands, and metal centers, recent efforts are actively concentrated on preparing stable (chemically and thermally) carbon structures with hierarchical porosity and active functional groups, such as conversion of the intrinsic metal centers to highly active catalytic oxides. The advantages of MDCs include controlled pore sizes and specific surface areas, which have been advancing in templating its intrinsic metal oxide and even extra functionality.

5.3 Heterogeneous Catalysis

The exploration of mesoporous non-siliceous hybrid materials as heterogeneous catalysts is of considerable interest due to their porosity which is favorable for the diffusion of reactants and products. It is possible to tailor the porous structures and functionality to yield chemo-, regio-, stereo- and/or enantioselectivity by creating an appropriate environment around the catalytic center in the restricted space available. Furthermore, owing to the homogeneous compositions of hybrid frameworks, a homogeneous distribution of active sites can thus be envisioned.

5.3.1 Pure Hybrid Framework

As an attractive alternative to reduce the consumption of fossil fuels, biodiesel is renewable and biodegradable and can be synthesized by transesterification of triglycerides or esterification of free fatty acids. Typically, the practical industrial production involves homogeneous acid (e.g., concentrated sulfuric acid) and alkaline catalysts, which are of serious environmental concern and pose difficulties in separation and purification of the target products. Recently, mesoporous tin phosphonate hybrid monolith was used to catalyze the esterification of long-chain fatty acids with methanol under mild conditions [75]. As an example, the catalyst showed excellent catalytic activity at room temperature and 94 % isolated yield was obtained for lauric acid. The catalytic activity showed negligible loss after five recycles. Supermicroporous iron phosphonate with interparticle mesoporosity found potential in transesterification reactions under solvent-free conditions [76]. The electrophilicity of the carbonyl carbon of the reactant ester group was the main driving force of the reaction. Negatively charged free P–OH in the porous framework could prevent molecules enriched with π-electron clouds from entering the porous channels. Thus, the π electrons were responsible for the minimal conversion of ethyl acrylate to the corresponding transesterified product in comparison with the other target esters. After the fifth cycle, the yield of methylcyanoacetate decreased slightly to 85.6 %, compared with 88.9 % in the first run, indicating the stability of the catalytic activity. The catalytic stability after multiple cycling could be due to the stability of Me–O–P bond, leading to the solid hybrid phosphonate frameworks and no leaching of active components during the reaction process.

Acid content is one of the key elements to determine catalytic activity and efficiency. Mesostructured zirconium organophosphonate, possessing a specific surface area of 702 m^2 g^{-1} and a uniform pore size of 3.6 nm, was synthesized with the assistance of C$_{16}$TABr using HEDP as coupling molecule [77]. The existence of defective P–OH was confirmed, showing an ion-exchange capacity of 1.65 mmol g^{-1}. The resultant hydroxyethylidene-bridged mesoporous zirconium phosphonate served as acid catalyst for the synthesis of methyl-2,3-O-isopropylidene-β-D-ribofuranoside from D-ribose, exhibiting high catalytic activity with rapid reaction rate (a product yield of 35.6 % after reaction at 70 °C for 3 h), which were comparable to the catalytic performance of liquid HCl (yield of 26.2 %) or commercial ion-exchange resin (yield of 33.0 %). The excellent catalytic activity could be attributable to the high surface area of the synthesized mesoporous zirconium phosphonates. The high specific surface area is beneficial for the distribution of the active sites and for their exposure so that they can readily be attached by the reactants, and the mesopores aid in the acceleration of mass transfer.

Nonetheless, it was difficult to achieve a high concentration of the desired metal phosphonates in the conventional synthesis methods, since the condensation between P–OH and metal ions during the preparation process often results in the extensive formation of P–O–M (M = Ti, Zr, V, Al, etc.) bonds. In order thus to increase the defective P–OH concentration in metal phosphates and phosphonates for the improvement of the H$^+$ exchange capacity, a series of alkyl amines were used as protecting groups during the condensation process, based on the reversible reaction between alkyl amines and P–OH groups in the phosphonic bridging molecules (Fig. 5.14) [64]. The alkyl amines first partially occupied the P–OH sites by acid–base reactions, followed by the condensation between the added alkoxides and residual P–OH and P = O groups. Extraction with HCl finally released the P–OH defects of the resultant solids, leading to a high H$^+$ exchange capacity and acid content. In the absence of amines, the P/Ti ratio reached a plateau of 1.35–1.51 when the added P/Ti ratio was larger than 1.75, due to the limit of coordination ability of Ti^{4+} ions with phosphonic acids. In the presence of amines, the P/Ti ratio of obtained

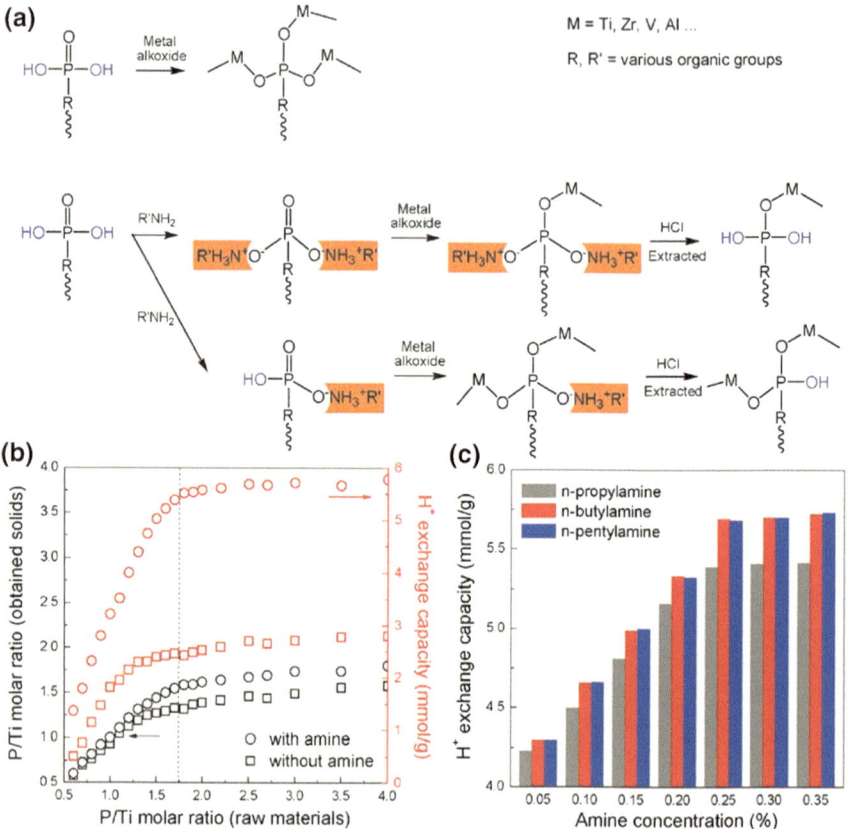

Fig. 5.14 Alkyl amine-assisted preparation of titanium phosphonates (**a**), the P/Ti molar ratios (**b**), and ion-exchange capacity (**c**) of the resultant titanium phosphonates. Reprinted with permission from Ref. [64]. Copyright 2011, Royal Society of Chemistry

solids exhibited a sharp initial rise and finally reached a plateau of 1.59–1.80, which was higher than without amines added. Correspondingly, a similar tendency was observed for the H^+ exchange capacity of the synthesized materials. The highest H^+ exchange capacities were confirmed as 2.44–2.79 and 5.51–5.80 mmol g^{-1} for the samples synthesized without and with amines assistance, respectively. A high yield of 48.7 % for methyl-2,3-O-isopropylidene-β-D-ribofuranoside was achieved. The product yield did not decrease even after 10 reuses.

Tin phosphonate hybrid with mesoscopic voids was synthesized using diphosphonic acid as spacers and employed as the catalyst for the polymerization of styrene to polystyrene in the absence of solvent and for the partial oxidation of styrene to phenylacetaldehyde and acetophenone in the presence of various aprotic solvents with dilute aqueous H_2O_2 as an initiator/oxidant [78]. The polymerization was completed at room temperature within 2–3 h, but the BET surface area of tetragonal tin phosphonate was relatively low (338 m^2 g^{-1}). Using C_{16}TABr as structure-directing agent, the surface area of tin phosphonate could be increased to $723 \ m^2 \ g^{-1}$, with the formation of micropores due to cross-linking of the ligand [79]. This hybrid demonstrated excellent catalytic activity in the direct one-pot oxidation of cyclohexanone to adipic acid using molecular oxygen under liquid phase conditions. The tin in the framework activated the molecular oxygen, helping to form the cyclic six-membered transition state, which further rearranged into a cyclic ester.

CO$_2$ can serve as C1 building block for various organic chemicals. One of the most promising reaction schemes currently seems to be the formation of cyclic carbonate via coupling of CO$_2$ and epoxides, which are useful as monomers, solvents, and pharmaceutical/fine chemical intermediates, and in biomedical applications [80]. Carboxylate-based MOFs have been extensively studied in this area [81–85]. Song et al. [81] reported the coupling reaction of CO$_2$ with propylene oxide to produce propylene carbonate catalyzed by MOF-5 in the presence of quaternary ammonium salts. The synergetic effect between MOF-5 and quaternary ammonium salts had excellent effect in promoting the reaction. The cycloaddition of CO$_2$ with epoxides is considered to be catalyzed by basicity and promoted by the Lewis acidic sites. On the basis of the intrinsic catalytic sites of the metal-connecting points (weak Lewis acid), the introduction of basic amino groups could lead to an enhanced catalytic performance [84]. Bifunctional hybrid catalysts containing moderate Lewis acidic and basic sites are preferred in the cycloaddition reactions. The bifunctionality can not only enhance the conversion and selectivity, but can also simplify the reaction conditions. The attempted bifunctionality can be obtained through judicious selection of precursors or through pre- and post-modification of the organic linkers. Up to now, reports concerning metal phosphonates for the coupling of CO$_2$ and epoxides have been relatively scarce. Since metal phosphonates show similar characteristics of composition and structure and higher stability compared to their carboxylate counterparts, it is meaningful to explore the catalytic activity of metal phosphonates in this burgeoning area.

Modified Fenton reactions have emerged as promising strategies for water treatment, especially for persistent and non-biodegradable pollutants. Generally, homogeneous Co^{2+}/peroxymonosulfate systems have been proven to be considerably efficient due to the powerful oxidizing ability of catalytically generated sulfate radicals toward the decomposition of organic molecules. On the other side, the introduction of multifarious transition metal centers in metal phosphonates can present distinct catalytic activities. For instance, cubic mesoporous titanium phosphonates showed superior photoactivity in degrading organic dyes under simulated solar light irradiation as compared with commercial P25 catalyst [11]. Effective catalytic hydrogenation of 4-nitrophenol to 4-aminophenol under ambient conditions could be achieved through the use of mesoporous nickel phosphate/phosphonate hybrid microspheres as the catalyst [86]. Mesoporous vanadium phosphonate material constructed from a dendritic tetraphosphonate could perform as an excellent catalyst for the aerobic oxidation of benzylic alcohols with high reactivity and shape selectivity [26]. It is reasonably speculated that porous cobalt phosphonate materials could fit the qualification of the Fenton reaction for oxidizing organic contaminants, which has been scarcely reported to the best of our knowledge. Organic–inorganic hybrid of cobalt phosphonate hollow nanostructured spheres was prepared in a water–ethanol system through a mild hydrothermal process in the absence of any templates using DTPMP as bridging molecule [87]. Cobalt phosphonate materials possessed amorphous frameworks with alternatively linked cobalt sites and organophosphonic bridging groups, providing abundant active sites for the catalytic oxidizing degradation of organic contaminants with the assistance of peroxymonosulfate under ambient conditions. The kinetic study showed that MB decomposition followed pseudo-first-order model with a heterogeneous reaction activation energy of 50.2 kJ mol^{-1}, and sulfate radicals were confirmed to be the active species. The experimental results suggested that cobalt phosphonate material could perform as an efficient heterogeneous catalyst for the degradation of organic contaminants, providing insights into the rational design and development of alternative catalysts for wastewater treatment.

In addition, the activity of hollow frameworks of mesoporous MOFs can be modified by tailoring the linkers to adjust the overall porosity or include chirality. Hwang et al. [88] proposed a way to selectively functionalize coordinatively unsaturated metal sites (CUS) in MIL-101 by attaching electron-rich functional groups onto unsaturated chromium sites. Trimeric chromium octahedral clusters in MIL-101 possess terminal water molecules, removable from the framework after vacuum treatment at 423 K for 12 h, thereby creating on the CUS Lewis acidic sites usable for the surface functionalization. The synthesis of the ethylenediamine-grafted MIL-101 (ED-MIL-101) was performed by coordinating ED to the dehydrated MIL-101 framework in toluene while heating under reflux. The amine-grafted MIL-101 exhibited a remarkably high activity in the base-catalyzed reaction and behaved as a size-selective catalyst. For example, the conversion of the condensation of benzaldehyde into trans-ethylcyanocinnamate on ED-MIL-101 was 97.1 %, with a selectivity of 99.1 %. Interestingly, ED-MIL-101 also revealed size dependency for the catalytic activities when the substituent groups of carbonyl compounds in the Knoevenagel condensation changed. Encapsulation of noble metals, such as palladium, over the amine-grafted MIL-101, has also been studied. Palladium-loaded APSMIL-101 and ED-MIL-101 have obviously high activities during the Heck reaction at 393 K that were comparable to those of a commercial Pd/C catalyst (1.09 wt% Pd) after a certain induction period (0.5–1 h), probably because of the slow diffusion of the reactants to reach accessible metal sites in the pores. Gascon et al. [89] further showed that MOFs with non-coordinated amino groups, IRMOF-3 and the amino-functionalized MIL-53, were stable solid basic catalysts in the Knoevenagel condensation of ethyl cyanoacetate and ethyl acetoacetate with benzaldehyde. The catalysts were stable under the studied reaction conditions and could be reused without significant loss in activity. The catalytic performance of IRMOF-3 in various solvents suggests that this open, accessible, and well-defined structure behaves more like a homogeneous basic catalyst, in contrast to other solid basic catalysts. MIL-101 could act as heterogeneous catalyst for the selective allylic oxidation of alkenes with *tert*-butyl hydroperoxide [90]. The selectivity toward α,β-unsaturated ketones reached 86–93 %. The temperature of the catalyst activation strongly affects the ketone yield.

MIL-101 is stable to chromium leaching, behaves as a true heterogeneous catalyst, can be easily recovered by filtration, and can be reused several times without loss of catalytic performance.

5.3.2 Post-functionalization for Catalysis

The homogeneously integrated organic functional moieties inside the hybrid framework permit the potential for post-functionalization to achieve further physicochemical characteristics, which is mainly based on the organic reactions. Chemically designed ordered mesoporous titanium phosphonate hybrid materials have exhibited the capacity to be functionalized by sulfation with chlorosulfonic acid ($ClSO_3H$) to form stable hydrosulfated esters (Fig. 5.15) [91]. The specific alkyl hydroxyl structure of the coupling molecule HEDP makes its sulfation facile. Approximately 2.69 and 3.93 mmol g^{-1} of H^+ were assigned to the grafted sulfonic groups and the defective P–OH from the hybrid framework, respectively. The acid strength revealed a Hammett indicator of $H_0 < -11.35$, indicative of a strong solid acid. It was also proven that the hydrosulfated groups remained at the pore walls even in hot water (up to 80 °C), which allowed the functionalized sample to be used as an ion exchanger and acid–base catalyst in room or low temperature reactions. For example, the sulfated materials could be used in the esterification of oleic acid and methanol under ambient temperature and pressure, giving a much higher conversion (87.3 %) than the unfunctionalized materials (4.9 %).

Transition metal-based catalysts are of great significance in sustainable environmental and energy chemistry. Inspired by the Langmuir adsorption behavior of metal ions onto porous phosphonates [11, 17], a step further would transform the metal ions into active components, which could have potential application in some catalytic reactions. To achieve a high dispersion of CuO nanoparticles, the

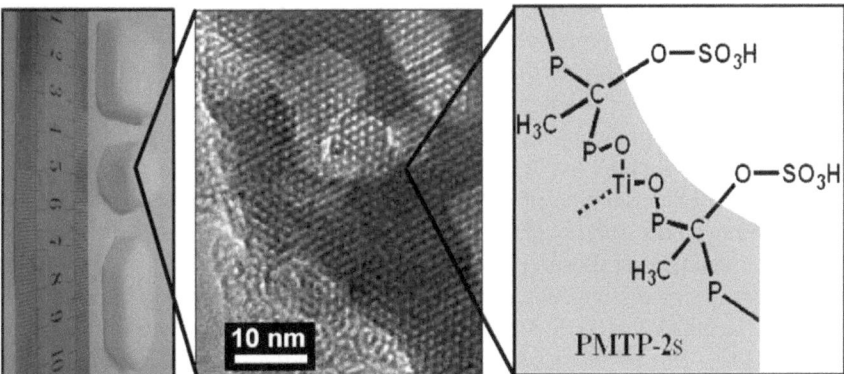

Fig. 5.15 Surface functionalization of monolithically ordered mesoporous titanium phosphonates prepared from HEDP using the $ClSO_3H$ treatment. Reprinted with permission from Ref. [91]. Copyright 2010, Royal Society of Chemistry

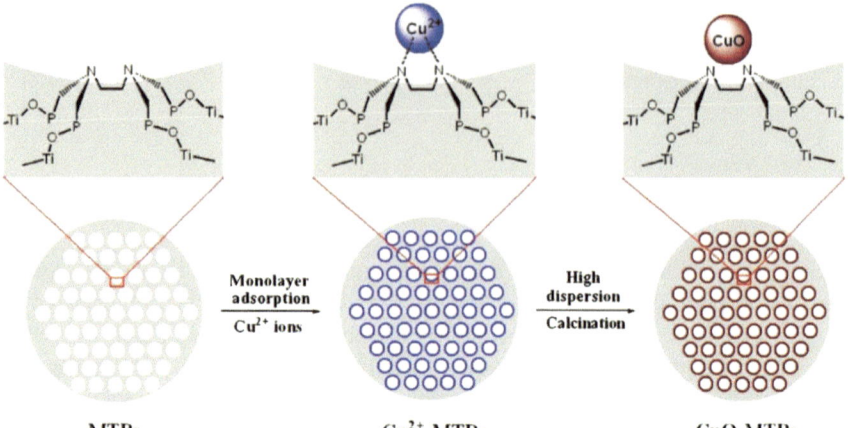

Fig. 5.16 Simulated formation process of CuO nanoparticles highly dispersed on metal phosphonate materials synthesized using EDTMP. Reprinted with permission from Ref. [92]. Copyright 2010, Royal Society of Chemistry

Cu^{2+} ions were firstly coordinated on the metal phosphonate material in the form of monolayers, and a subsequent calcination at 450 °C generated the highly dispersed CuO active components (Fig. 5.16) [92]. Since the PMTP-1 microspheres were thermally stable to 450 °C, the high-temperature calcination could achieve a high dispersion of CuO while maintaining the mesoporous hybrid framework. The density and distribution of the surface organic functional groups could be tuned, allowing for an indirect adjustment of the dispersion of the Cu^{2+} and the final CuO loading amounts. A main reduction peak at 217 °C was observed in the temperature-programmed reduction (H_2-TPR) analysis of the synthesized catalysts, which was lower than for pure CuO and CuO catalysts supported on inorganic metal phosphates without organic ligands (252 °C). It is commonly accepted that a high dispersion of active components on the supports can contribute toward improving the catalytic oxidation performance [93, 94]. The oxidation of toxic CO was selected as the probe reaction, and the catalytic activity of the synthesized supported catalyst was higher than those materials with the same CuO content but prepared by the conventional impregnation methods. Moreover, the synthesized catalyst showed a significant stability for low-temperature CO catalytic oxidation.

Noble metals, such as Au, Ag, Pt, and Pd, have been known for their catalytic performances. Encapsulation of noble metal nanoparticles inside the metal phosphonate frameworks may extend their applications in catalysis and energy conversion by catalytic spillover [95, 96]. Interestingly, noble metal nanoparticles with different size regimes could be made through different reduction methods, reduction in ethanol (10–15 nm), and at elevated temperature under hydrogen (2–4 nm) [93]. This provides a simple way to control the size of the loaded active components. Canivet et al. [97] developed one-pot post-synthetic grafting of a nickel-based organometallic catalyst within a MOF framework under mild conditions, preventing the interactions of the organic graft with the metal nodes and the encapsulation of metal particles. The imine condensation occured in the presence of the Ni(PyCHO)Cl₂ (PyCHO = 2-pyridine carboxaldehyde) methanolic solution to directly generate the diimino nickel complex anchored into the MOF. This method allowed the rapid preparation of a MOF-based catalyst whose activity and selectivity were demonstrated for the selective ethylene dimerization to give the corresponding alpha olefin (1-butene) in liquid phase. Leaching test showed that the reaction did not proceed any longer if the filtrated catalytic solution is again put under catalytic conditions in the presence of Al-based cocatalyst.

MOF-based supported heterogeneous catalysis has attracted increasing research attention because the nanopores of MOFs can not only serve as templates for synthesizing monodisperse active species but also provide well-defined microenvironments that could induce selectivity control on the encapsulated active species in catalytic reactions. The loading of metal nanoparticles/nanoclusters into MOFs was first realized by chemical vapor deposition using volatile organometallic precursors in the gas phase [98, 99]. Moreover, combination of active metal NPs with the functionalities within the host, multifunctional catalysts capable of promoting different reactions, or one-pot cascade reactions can thus be realized. A preferred system for catalyst study is Pd@mesoMOFs [100, 101], and other systems including Au@mesoMOFs [102] and Ni@mesoMOFs [103] are also attractive. The catalytic reactions are mostly focused on oxidation of alcohols, hydrogenation reaction, and C–C coupling reaction. For instance, Li and He et al. reported the one-pot cascade synthesis of methyl isobutyl ketone (MIBK) from acetone catalyzed by Pd@MIL-101(Cr), which was obtained by a simple impregnation method [101]. MIBK is manufactured via a three-step process involving condensation, dehydration, and hydrogenation, wherein 0.11–0.34 wt% Pd loading was found to be suitable for the desired selectivity and conversion, probably due to the monodispersion of NPs throughout the pores. Kim and coworkers successfully prepared Ni nanoparticles embedded inside a mesoporous MOF (MesoMOF-1) using gas-phase loading and subsequent reduction (Fig. 5.17) [103]. The resulting Ni@MesoMOF-1 could further act as a catalyst for hydrogenolysis of nitrobenzene or hydrogenation of styrene with good activity.

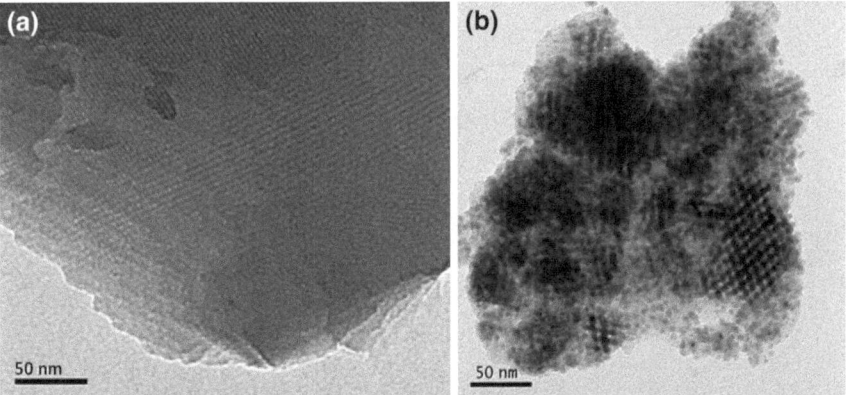

Fig. 5.17 TEM images of **a** MesoMOF-1 and **b** Ni@MesoMOF-1. Lattice fringes are observed in both (**a**) and (**b**), indicating Ni nanoparticles are aligned with a long-range order. Reprinted with permission from Ref. [103]. Copyright 2010, Royal Society of Chemistry

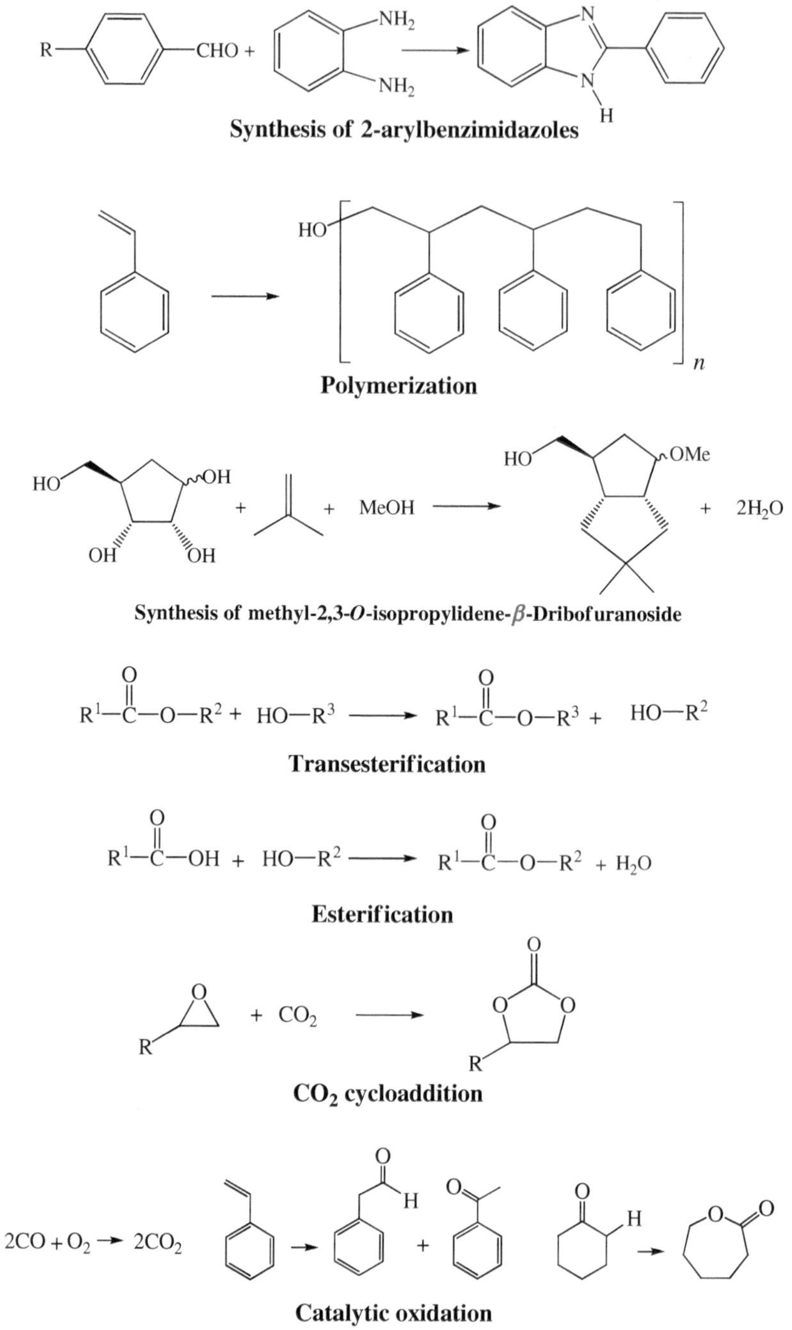

Synthesis of 2-arylbenzimidazoles

Polymerization

Synthesis of methyl-2,3-*O*-isopropylidene-*β*-Dribofuranoside

$$R^1-\overset{\overset{\displaystyle O}{\|}}{C}-O-R^2 + HO-R^3 \longrightarrow R^1-\overset{\overset{\displaystyle O}{\|}}{C}-O-R^3 + HO-R^2$$

Transesterification

$$R^1-\overset{\overset{\displaystyle O}{\|}}{C}-OH + HO-R^2 \longrightarrow R^1-\overset{\overset{\displaystyle O}{\|}}{C}-O-R^2 + H_2O$$

Esterification

CO₂ cycloaddition

$$2CO + O_2 \longrightarrow 2CO_2$$

Catalytic oxidation

Fig. 5.18 Summary of some emerging and potential catalytic applications of mesoporous non-siliceous hybrids and their related composites

Catalytic reactions are indeed surface interaction processes, where the metal joints and organic bridging groups perform as active sites. The special compositions and a variety of structural features can engender catalytic activity (Fig. 5.18). Post-modification can mainly be based on the organic motifs inside the framework, which can create novel physicochemical properties, such as acidity/alkalinity, hydrophobicity/hydrophilicity, and chirality. However, there are still issues concerning thermal stability and chemical robustness, due to the nature of the components of the materials. Catalysis supported by mesoporous non-siliceous hybrid materials is still in its infancy.

5.4 Biomaterials

Numerous applications of nanomaterials emerge in biology owing to nanosizes matching well within the dimensions of organisms. Low-dimensional nanomaterials such as nanospheres and nanoparticles have found application in biomedicines and biotechnologies. Furthermore, mesoporous hybrid materials are remarkably attractive due to the synergistic roles of both interactions with biological molecules at the organic–inorganic surfaces and confinement in regular mesopores.

5.4.1 Biomolecule Adsorption and Separation

Immobilization of enzymes on solid supports can improve enzyme stability, facilitate separation and recycling, and maintain the catalytic activity and selectivity [104]. Classical mesoporous silicas usually suffer from the easy leaching of immobilized enzyme molecules due to the lack of interactions between enzymes and host materials, which results in the loss of catalytic activity upon multiple uses reversely. Microperoxidase-11 (MP-11) has dimensions of about $3.3 \times 1.7 \times 1.1$ nm. The pore sizes of Tb-mesoMOF dominantly distributed around 3.0 and 4.1 nm in addition to a small portion of micropore size around 0.9 nm. Organic components in the hybrid materials (MP-11) could be successfully immobilized in mesoporous MOFs containing nanoscopic cages of around 4.0 nm under the drive of host–guest hydrophobic interactions [105]. The corresponding loading amount could reach 19.1 mmol g^{-1}. Accordingly, as compared with the mesoporous silica counterpart, the resulting enzyme-loaded MOFs exhibited superior enzymatic catalysis performances and reusability for polyphenol oxidation in the presence of hydrogen peroxide than the mesoporous silica counterparts (Fig. 5.19).

Mesoporous zirconium organophosphonates using 1-phosphomethylproline (H$_3$PMP) as the bridging molecule possess tunable mesopores, high surface area, and large pore volume, exhibiting high adsorption capacity and adsorption rates for enzymes [106]. For lysozyme (Lz) adsorption, the adsorption equilibrium was

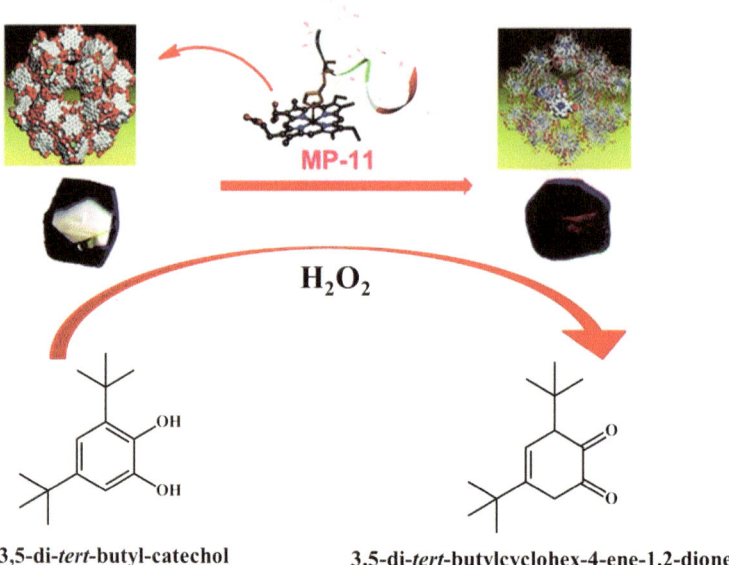

3,5-di-*tert*-butyl-catechol   **3,5-di-*tert*-butylcyclohex-4-ene-1,2-dione**

Fig. 5.19 Immobilization of MP-11 in mesoporous MOFs for catalytic oxidation of 3,5-di-tert-butylcatechol to o-quinone. Reprinted with permission from Ref. [105]. Copyright 2011, American Chemical Society

reached within 30 min. The adsorption capacity for Lz and papain was as high as 438 and 297 mg g^{-1}, respectively. Furthermore, Lz loaded on mesoporous zirconium phosphonates retained a structural conformation similar to its free state, suggesting that no denaturation of Lz occurred during the adsorption process. No leaching of Lz from the solid was observed when shaking the Lz-loaded solid in a buffer solution. The loading of biomolecules into the porous phosphonate hybrid networks is directly correlated with the strength of host–guest interactions [107], surface area, and pore size [105, 106]. Correspondingly, separation of biomolecules can be feasibly realized through utilizing the targeted phosphonic bridging groups and adjusting the porosity of the phosphonate materials.

The controllable adsorption and separation of proteins are indispensable for the application of biosensors, biocatalysts, and disease diagnostics. Size-selective adsorption of guest protein molecules, which mainly depends on the porous properties of sorbents, has attracted tremendous research interest due to the feasibility to adjust the porosity of the host solid materials. Although usual porous sorbents have good capacities toward adsorbates, they still confront the predicament in separating proteins from each other. Hollow manganese phosphonate microspheres (HMPM) with hierarchical porosity showed size selectivity toward Cytochrome C (Cyt C, 12,400 Da, 2.6 × 3.2 × 3.3 nm^3) and the protein bovine serum albumin (BSA, 66,400 Da, 5.0 × 7.0 × 7.0 nm^3) [108]. The result of simultaneous adsorption on a manganese phosphonate spherical hybrid is shown in Fig. 5.20.

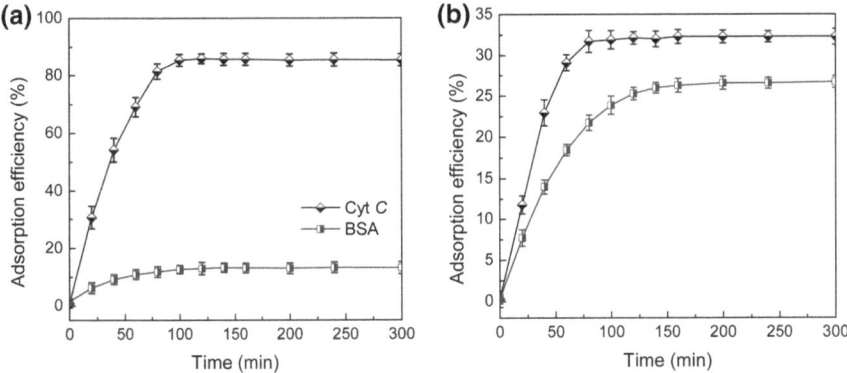

Fig. 5.20 Time-dependent adsorption curves of Cyt C and BSA for the pristine HMPM (**a**) and the HMPM material after a ball-milling treatment (**b**). Reprinted with permission from Ref. [108]. Copyright 2014, Royal Society of Chemistry

Cyt C adsorption increased sharply at the initial contact time and slowed down gradually until adsorption equilibrium was reached, demonstrating an adsorption efficiency of 85.7 %. In contrast, the HMPM hybrid showed much lower adsorption capability for BSA (13.1 %). This might have resulted from the porous hierarchy of HMPM. The mesostructured pores with small pore size were distributed in the shell sections of manganese phosphonate microspheres, which would permit the penetration of small molecules (Cyt C) through the adsorbent and block BSA molecules with larger size. Thus, the analogous "semipermeable membrane" effect would favor the separation of proteins with different sizes. Noticeably, the microspherical morphology with porous hierarchy was destroyed by a ball-milling technique before adsorbing proteins, leading to a slight decrease of specific surface area to 48 m^2 g^{-1} and a wide pore size distribution from 0 to 40 nm. As seen in Fig. 5.20b, both the protein molecules can be adsorbed and the separation goal cannot be achieved. Moreover, the resultant adsorption ability of Cyt C is much lower than that of HMPM, which may be due to the existence of competitive adsorption of the two proteins on the sorbent surface. Therefore, such a good selectivity is mainly attributable to the peculiar porosity of the manganese phosphonate microspheres.

5.4.2 Drug Delivery

The storage capacity and release of drug in porous host materials are governed by various factors such as pore size, shape, connectivity, and host affinity. As to traditional porous materials including silica and polymeric matrixes, drug loading capacity is usually not sufficiently high and encapsulated drug is difficult to be released specifically. Porous materials with large volumes and regular structures are desired to realize a high loading and a controlled release. Férey et al. tested the abilities of MIL-100(Cr) and MIL-101(Cr) for the delivery of ibuprofen, showing remarkable adsorption with 0.347 and 1.376 g g^{-1}, respectively [109]. The complete release of ibuprofen was achieved for MIL-100(Cr) and MIL-101(Cr) after 3 and 6 days, respectively. Compared with MCM-41, MIL-101(Cr) demonstrated a four times higher loading capacity and much longer release time (2 days for MCM-41), which was probably due to the stronger interaction between ibuprofen and the MIL-101(Cr) framework (π–π and acid–base interactions). However, chromium is toxic, prohibiting further clinical applications. The biocompatible nanoscale Fe-MIL-100, along with microporous iron(III) carboxylate, was prepared by Férey et al. for drug delivery and imaging to reconcile the cytotoxicity and high drug loading capacity [110]. The low toxicity was confirmed by the reversible weight increase and return to normality after injection in 1–3 months as well as the absence of immune or inflammatory reactions. The antitumoural drug busulfan (Bu) could be loaded into Fe-MIL-100, and the same activity was obtained for the entrapped Bu as free Bu due to entrapped Bu in its molecular form within the pores.

Controlled drug delivery and release technology offer numerous advantages in comparison with conventional dosage forms including improved efficacy, reduced toxicity, and improved patient compliance and convenience. This process mainly depends on the variations of pH, light, redox potential, and temperature. The designed delivery systems of "molecular lock" are able to selectively release the entrapped guests. The incorporation of 1,4-bis(phosphomethyl)piperazine (BPMP) introduced pH sensitivity into the metal phosphonate hybrid network (Fig. 5.21) [111]. The pH sensitivity was derived from the reversible protonation under acidic conditions and deprotonation with weakly basic piperazines under different pH

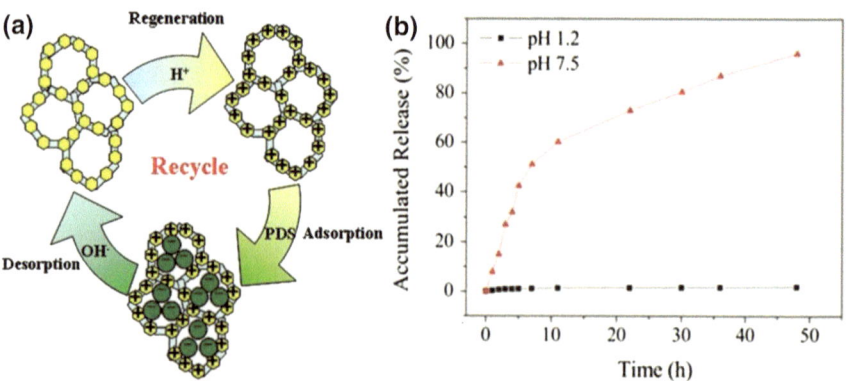

Fig. 5.21 **a** A recycle process of pH-sensitive ZrBPMP materials: adsorption, desorption, and regeneration. **b** The PDS release profiles of PDS-loaded mesoporous zirconium phosphonates in simulated gastric fluid (pH = 1.2) and intestinal fluid (pH = 7.5). Reprinted with permission from Ref. [111]. Copyright 2010, Royal Society of Chemisty

conditions, thereby endowing mesoporous zirconium phosphonates with reversible cationic–neutral surface properties. The designed delivery systems of "molecular lock" are able to selectively release the entrapped guests. For instance, the negatively charged PDS (a photosensitizer of sulfonated phthalocyanine for photodynamic therapy of tumors) could then be adsorbed or released through strong electrostatic interaction according to the pH conditions. The integration of H_3PMP and BPMP would lead to phosphonate hybrids with bifunctionality, pH sensitivity, and functionalizability [111]. The reversible protonation–deprotonation of L-proline groups of H_3PMP and piperazine groups of BPMP on the mesoporous walls under different pH values (pH sensitivity) as well as the further functionalization with cell-penetrating peptides via the carboxyl in L-proline group of H3PMP on outer surface (functionalizability) endowed the materials with pH-controllable release function and high cell penetration capability [112]. Thus, a time- and pH-controlled oral colon-targeted nucleic acid delivery system was developed. Using salmon sperm DNA as model nucleic acid allowed it to remain intact during delivery. The penetration capability through biomembranes was enhanced through further functionalization with a cell-penetrating peptide of octaarginine.

5.4.3 Biosensors

Nanoscale Fe-MIL-101_NH_2 presented an alternative way for drug delivery and imaging via PSM (Fig. 5.22) [113]. The presence of amino groups on the particles allowed for covalent attachment of biologically relevant cargoes by PSM. An optical imaging contrast agent was firstly loaded by treating the NPs with BODIY-Br (1,3,5,7-tetramethyl-4,4-difluoro-8-bromomethyl-4-bora-3a,4a-diazasindacence) in THF at room temperature. The BODIPY loading was determined to be 5.6–11.6 wt%. Confocal microscope images of the BODIPY-loaded particles with HT-29 cells demonstrated fluorescence, which was an indication for the penetration of the particles through cell membrane and release of the fluorescent cargoes. The ethoxysuccinato-cisplatin (ESCP), a prodrug of cisplatin, was also loaded by treating the NPs with ESCP, which was first activated by 1,1-carbonyldiimidazole in DMF at room temperature. Thus, the NH_2-functionalized MIL-101(Fe) NPs provided an efficient platform for delivering the ESCP prodrug with an overall payload of 12.8 wt%. However, these NPs were not stable in PBS buffer at 37 °C and experienced rapid degradation. To improve the control of degradation, these particles were coated with a thin silica layer, resulting in a longer half-life of 16 and 14 h for the BODY- and ESCP-loaded particles, respectively, in PBS buffer at 37 °C compared to 2.5 and 1.2 h of the uncoated particles. Functionalization of silica-coated particles with cyclic peptide c(RGDfk) showed that these NPs had cytotoxicity comparable to that of cisplatin when treated against HT-29 cells. The work affords a valid approach for the design of a wide range of nanomaterials for imaging and therapeutic applications.

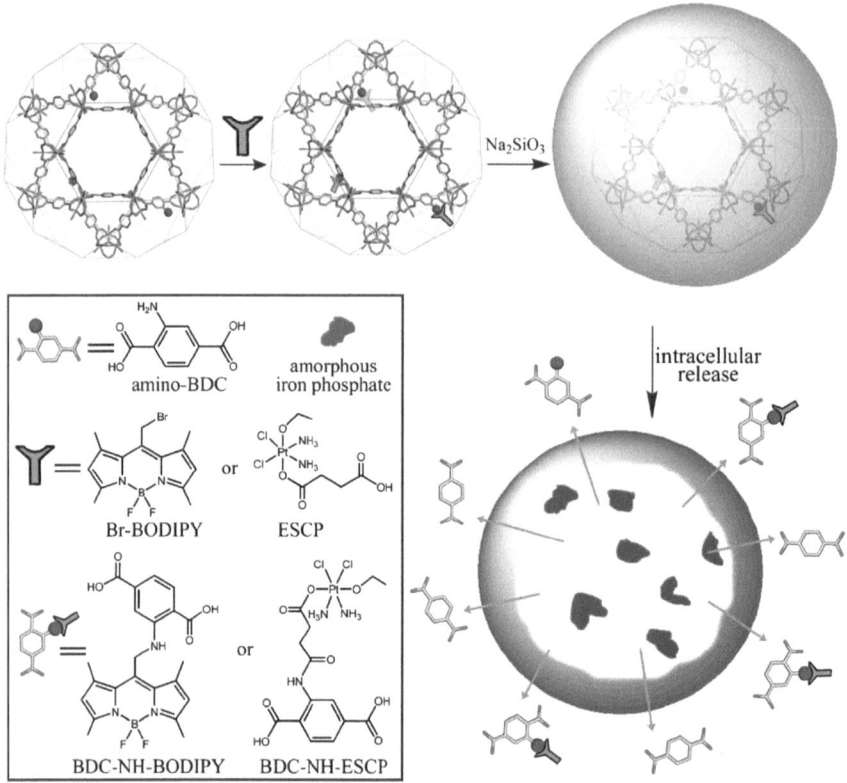

Fig. 5.22 Schematic representation of covalent attachment of BODIPY-Br and ESCP, further coating with silica and final release as imaging contrast agent and anticancer drug. Reprinted with permission from Ref. [113]. Copyright 2009, American Chemical Society

Numerous studies have indicated that nanoparticle-based therapeutics and diagnostic agents show enhanced efficacy and reduced side effects, due to their unique physicochemical properties [114, 115]. The vast majority of nanocarriers can be classified into two categories: either purely inorganic (e.g., quantum dots) or purely organic (e.g., liposomes). Noticeably, nanosized metal phosphonate hybrids have the potential to combine attractive characteristics of both inorganic and organic nanocarriers including robust particle morphologies, compositional and structural diversity, biocompatibility, and bioactivity, to provide a unique platform for delivering agents, therapeutics, and biosensing [116]. Surface modification of iron oxide nanoparticles by phosphonates has a wealth of applications including magnetic resonance imaging (MRI), drug delivery, and hyperthermia for cancer treatment [117–119]. Lartigue et al. [119] reported the modification of iron oxide nanoparticles with carbohydrates derivatized by phosphonate groups. The magnetic, hyperthermal, and relaxometric properties of the phosphonated nanoparticles made them promising candidates for MRI imaging and hyperthermia. On the basis of poly(quaternary ammonium) brushes grown by atom transfer radical polymerization using an initiator grafted via a phosphonate group to the surface of magnetite nanoparticles [120], recyclable antibacterial magnetic nanoparticles were successfully synthesized. Given the convenience of separation of the nanoparticles from the bacterial culture tests using an external magnetic field, the resultant nanoparticles presented high antibacterial activity against *E. coli* even after eight exposure tests. When cyclodextrin groups were attached to magnetite nanoparticles using a phosphonic linkage [121], the anchored cyclodextrin formed inclusion complexes with diclofenac sodium salt, a non-steroidal anti-inflammatory drug, demonstrating the potential for targeted drug delivery.

In the past few years, some implant semiconductor biomaterials functionalized by phosphonic acids, such as In_2O_3 and TiO_2, have been investigated for biosensor applications. In_2O_3 nanowires were first grafted with 3-phosphonopropionic acid, and then, the terminal carboxylic acid groups were activated by EDC–NHS aqueous solution [122], resulting in a nanowire surface reactive toward the amine groups present on antibodies. After passivation with an amphipathic polymer (Tween 20), the resultant sensors were found to be capable of performing rapid, label-free, electrical detection of cancer biomarkers directly from human whole blood collected by a finger prick. However, up to now, detection and treatment of organism diseases are two consecutive and inseparable processes in clinical diagnostics and medicine, but their academic studies are often isolated from each other. It is still challenging and significant to design a "diagnospy" carrier that combines the functions of biomolecule quantitative detection and bioresponsive drug controlled release [123, 124]. An interesting study pioneered by Li et al. [125] was to intentionally design a smart system on the basis of hybrid phosphonate–TiO_2 mesoporous nanostructures capped with fluorescein labeled oligonucleotides, which could realize simultaneous and highly efficient biomolecule sensing and controlled drug release (Fig. 5.23). The incorporation of phosphonate could shift the absorption edge of titania to the visible light range and introduce positively charged amino groups to interact with negatively charged fluorescein labeled oligonucleotides, resulting in the closing of the mesopores and the fluorescence quenching of fluorescein at the same time. The further addition of complementary single DNA strands or protein target led to the displacement of the capped DNA due to hybridization or protein–aptamer reactions. Correspondingly, the pores were opened, causing the release of entrapped drugs as well as the restoration of dye fluorescence. Moreover, target concentration-dependent fluorescent signal response could be used to monitor treatment effects in real time, thus providing proof for determining drug dose or adjusting the treatment program. The luminescence intensity linearly increased with the increasing of thrombin concentration, until a plateau was reached. There was a good linearity relationship between the $(F/F_0 - 1)$ value and thrombin concentration increasing from 5 to 175 nm with the correlation coefficient of 0.996. The limit of detection (LOD) was 2.3 nm. Interference experiments exhibited that human serum albumin, collagenase, lysozyme, cytochrome c, hemoglobin, and trypsin presented much lower fluorescence intensity restoration and drug release capacities than that of human thrombin due to the almost unopened aptamer-capped mesopores. This mesoporous hybrid system provides a

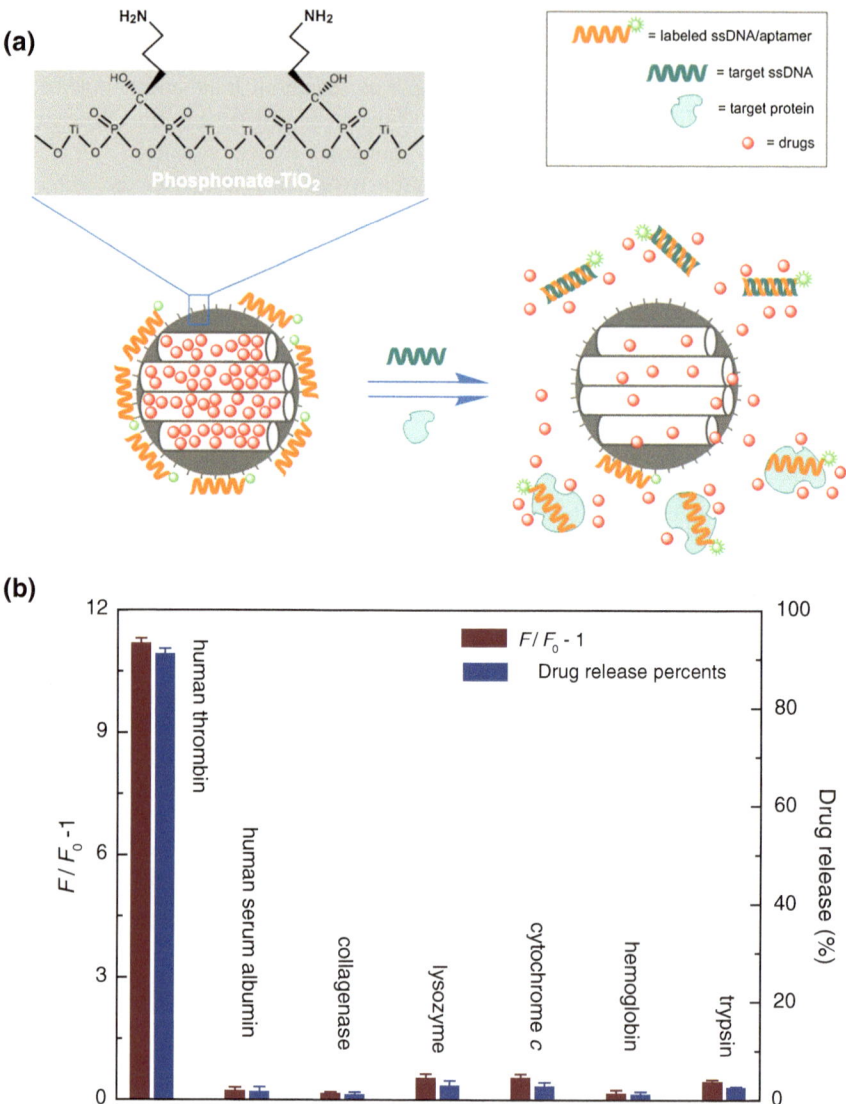

Fig. 5.23 **a** Schematic illustration of bioresponsive detection and drug controlled release system based on phosphonate–TiO$_2$ hybrid material. **b** Bioresponsive sensing and drug release selectivity of FAM–O–PTi system among different proteins. Reprinted with permission from Ref. [125]. Copyright 2013, Royal Society of Chemistry

novel perception to utilize non-siliceous hybrid materials as the supports in sensing and control release applications.

Mesoporous non-siliceous hybrid materials can perform as host materials in the fields of biosensing and biotechnology owing to their well-defined porosity, low biotoxicity, and capacity of incorporation of biogroups. Introduction of specific organic bridging groups could result in distinct biomimic performance. However, many practical detecting or sensing applications require extraordinarily high sensitivities. Optical sensors are molecular receptors whose optical properties can be changed upon binding to specific guests. Optical sensing and imaging systems have been intensively investigated for their capability of providing high sensitivity, fast, and easy detection processing, biocompatibility, and adaptability to a wide variety of conditions [126]. Since lanthanide-based hybrids are photoluminescent materials with ease of functionalization, they are a promising class of materials for applications in sensing and optical imaging. Mesoporous cerium phosphonate nanostructured hybrid spheres are prepared with the assistance of C_{16}TABr while using EDTMP as the coupling molecule [127]. The resulting hybrid is constructed from the cerium phosphonate nanoparticles, accompanied by high specific surface area of 455 m^2 g^{-1}. The uniform incorporation of rare earth element cerium and organophosphonic functionalities endows mesoporous cerium phosphonate with excellent fluorescence properties for the development of an optical sensor for selective Hg^{2+} detection on the basis of the fluorescence-quenching mechanism. The signal response of mesoporous cerium phosphonate against the Hg^{2+} concentration is linear over the range from 0.05 to 1.5 μmol L^{-1}, giving a LOD of 16 nmol L^{-1} (at a signal-to-noise ratio of 3) (Fig. 5.24). Most of the common physiologically relevant cations and anions did not interfere with the detection of Hg^{2+}. Although lanthanide-based MOFs with valuable luminescent properties, which can be defined as crystalline organic–inorganic hybrids to some extent, have been gradually utilized as optical sensing materials [128–130], nevertheless, there are still some knotty problems. The major one is the insufficient water solubility that restricts the further uses in biologic systems, and the emission band usually contains multipeaks, reducing the monochromaticity and measurement precision. Hence, the present mesoporous cerium phosphonate hybrid nanostructured spheres

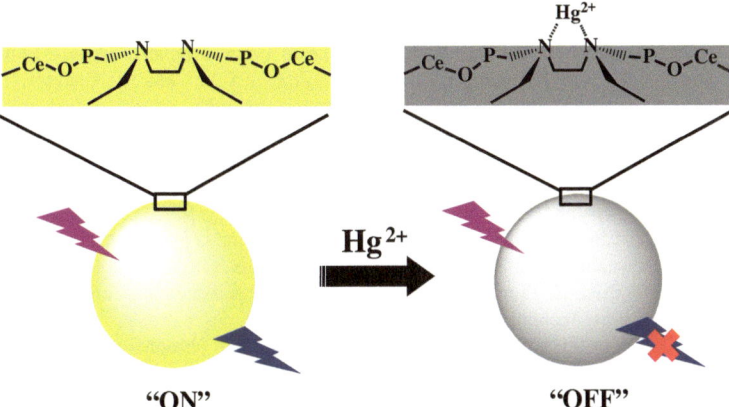

Fig. 5.24 Scheme of Hg^{2+} detection by mesoporous cerium phosphonate nanostructured spheres based on fluorescence-quenching mechanism. Reprinted with permission from Ref. [127]. Copyright 2014, American Chemical Society

with well-defined porosity and good dispersity in water hold a promising potential for practical biosensing applications.

References

1. H. Furukawa, N. Ko, Y.B. Go, N. Aratani, S.B. Choi, E. Choi, A. Özgür Yazaydin, R.Q. Snurr, M. O'Keeffe, J. Kim, O.M. Yaghi, Ultrahigh porosity in metal–organic frameworks. Science **329**, 424–428 (2010)
2. Q.R. Fang, D.Q. Yuan, J. Sculley, W.G. Lu, H.C. Zhou, A novel MOF with mesoporous cages for kinetic trapping of hydrogen. Chem. Commun. **48**, 254–256 (2012)
3. B. Mu, P.M. Schoenecker, K.S. Walton, Gas adsorption study on mesoporous metal–organic framework UMCM-1. J. Phys. Chem. C **114**, 6464–6471 (2010)
4. L. Liu, Q.F. Deng, X.X. Hou, Z.Y. Yuan, User-friendly synthesis of nitrogen-containing polymer and microporous carbon spheres for efficient CO_2 capture. J. Mater. Chem. **22**, 15540–15548 (2012)
5. L. Liu, Q.F. Deng, T.Y. Ma, X.Z. Lin, X.X. Hou, Y.P. Liu, Z.Y. Yuan, Ordered mesoporous carbons: citric acid-catalyzed synthesis, nitrogen doping and CO_2 capture. J. Mater. Chem. **21**, 16001–16009 (2011)
6. R. Serna-Guerrero, E. Dána, A. Sayari, New insights into the interactions of CO_2 with amine-functionalized silica. Ind. Eng. Chem. Res. **47**, 9406–9412 (2008)
7. Z.X. Wu, N. Hao, G.K. Xiao, L.Y. Liu, P. Webley, D.Y. Zhao, One-pot generation of mesoporous carbon supported nanocrystalline calcium oxides capable of efficient CO_2 capture over a wide range of temperatures. Phys. Chem. Chem. Phys. **13**, 2495–2503 (2011)
8. T.Y. Ma, Z.Y. Yuan, Organic-additive-assisted synthesis of hierarchically meso-/macroporous titanium phosphonates. Eur. J. Inorg. Chem. **19**, 2941–2948 (2010)
9. Y.P. Zhu, T.Z. Ren, Z.Y. Yuan, Mesoporous non-siliceous inorganic–organic hybrids: a promising platform for designing multifunctional materials. New J. Chem. **38**, 1905–1922 (2014)
10. T.Y. Ma, X.Z. Lin, X.J. Zhang, Z.Y. Yuan, High surface area titanium phosphonate materials with hierarchical porosity for multi-phase adsorption. New J. Chem. **34**, 1209–1216 (2010)
11. T.Y. Ma, X.Z. Lin, Z.Y. Yuan, Cubic mesoporous titanium phosphonates with multifunctionality. Chem. Eur. J. **16**, 8487–8494 (2010)
12. G.P. Knowles, S.W. Delaney, A.L. Chaffee, Diethylenetriamine[propyl(silyl)]-functionalized (DT) mesoporous silicas as CO_2 adsorbents. Ind. Eng. Chem. Res. **45**, 2626–2633 (2006)
13. X.J. Wang, P.Z. Li, Y.F. Chen, Q. Zhang, H.C. Zhang, X.X. Chan, R. Ganguly, Y.X. Li, J.W. Jiang, Y.L. Zhao, A rationally designed nitrogen-rich metal–organic framework and its exceptionally high CO_2 and H_2 Uptake Capability. Sci. Report **3**, 1–5 (2013)
14. L.F. Song, J. Zhang, L.X. Sun, F. Xu, F. Li, H.Z. Zhang, X.L. Si, C.L. Jiao, Z.B. Li, S. Liu, Y.L. Liu, H.Y. Zhou, D.L. Sun, Y. Du, Z. Cao, Z. Gabelica, Mesoporous metal–organic frameworks: design and applications. Energy Environ. Sci. **5**, 7508–7520 (2012)
15. T.Y. Ma, X.Z. Lin, X.J. Zhang, Z.Y. Yuan, Hierarchical mesostructured titanium phosphonates with unusual uniform lines of macropores. Nanoscale **3**, 1690–1696 (2011)
16. M.M. Maroto-Valer, Z. Tang, Y. Zhang, CO_2 capture by activated and impregnated anthracites. Fuel Process. Technol. **86**, 1487–1502 (2005)
17. T.Y. Ma, X.Z. Lin, Z.Y. Yuan, Periodic mesoporous titanium phosphonate hybrid materials. J. Mater. Chem. **20**, 7406–7415 (2010)
18. K.Z. Hossain, L. Mercier, Intraframework metal ion adsorption in ligand-functionalized mesoporous silica. Adv. Mater. **14**, 1053–1056 (2002)
19. T.Z. Ren, X.H. Zhu, T.Y. Ma, Z.Y. Yuan, Adsorption of methylene blue from aqueous solution by periodic mesoporous titanium phosphonate materials. Adsorpt. Sci. Technol. **31**, 535–548 (2013)

20. J. Kim, R.J. Desch, S.W. Thiel, V.V. Guliants, N.G. Pinto, Energetics of biomolecule adsorption on mesostructured cellular foam silica. Micropor. Mesopor. Mater. **170**, 95–104 (2013)
21. Y.P. Zhu, T.Y. Ma, Y.L. Liu, T.Z. Ren, Z.Y. Yuan, Metal phosphonate hybrid materials: from densely layered to hierarchically nanoporous structures. Inorg. Chem. Front. **1**, 360–383 (2014)
22. F.N. Shi, L. Cunha-Silva, R.A.S. Ferreira, L. Mafra, T. Trindade, L.D. Carlos, F.A.A. Paz, J. Rocha, Interconvertable modular framework and layered lanthanide(III)-etidronic acid coordination polymers. J. Am. Chem. Soc. **130**, 150–167 (2008)
23. Z.Y. Yuan, B.L. Su, Insights into hierarchically meso-macroporous structured materials. J. Mater. Chem. **16**, 663–677 (2006)
24. T. Salesch, S. Bachmann, S. Brugger, R. Rabelo-Schaefer, K. Albert, S. Steinbrecher, E. Plies, A. Mehdi, C. Reyé, R.J.P. Corriu, E. Lindner, New inorganic-organic hybrid materials for HPLC separation obtained by direct synthesis in the presence of a surfactant. Adv. Funct. Mater. **13**, 134–142 (2002)
25. T.Y. Ma, H. Li, A.N. Tang, Z.Y. Yuan, Ordered, mesoporous metal phosphonate materials with microporous crystalline walls for selective separation techniques. Small **7**, 1827–1837 (2011)
26. M. Vasylyev, R. Neumann, Preparation, characterization, and catalytic aerobic oxidation by a vanadium phosphonate mesoporous material constructed from a dendritic tetraphosphonate. Chem. Mater. **18**, 2781–2783 (2006)
27. M. Vasylyev, E.J. Wachtel, R. Popovitz-Biro, R. Neumann, Titanium phosphonate porous materials constructed from dendritic tetraphosphonates. Chem. Eur. J. **12**, 3507–3514 (2006)
28. A. Clearfield, Unconventional metal organic frameworks: porous cross-linked phosphonates. Dalton Trans. **28**(44), 6089–6102 (2008)
29. K.J. Gagnon, H.P. Perry, A. Clearfield, Conventional and unconventional metal–organic frameworks based on phosphonate ligands: MOFs and UMOFs. Chem. Rev. **112**, 1034–1054 (2012)
30. H.L. Jiang, Y. Tatsu, Z.H. Lu, Q. Xu, Non-, micro-, and mesoporous metal–organic framework isomers: reversible transformation, fluorescence sensing, and large molecule separation. J. Am. Chem. Soc. **132**, 5586–5587 (2010)
31. T.Y. Ma, X.J. Zhang, G.S. Shao, J.L. Cao, Z.Y. Yuan, Ordered macroporous titanium phosphonate materials: synthesis, photocatalytic activity, and heavy metal ion adsorption. J. Phys. Chem. C **112**, 3090–3096 (2008)
32. T.Y. Ma, X.J. Zhang, Z.Y. Yuan, Hierarchically meso-/macroporous titanium tetraphosphonate materials: synthesis, photocatalytic activity and heavy metal ion adsorption. Micropor. Mesopor. Mater. **123**, 234–242 (2009)
33. Y.P. Zhu, M. Li, Y.L. Liu, T.Z. Ren, Z.Y. Yuan, Carbon-doped ZnO hybridized homogeneously with graphitic carbon nitride nanocomposites for photocatalysis. J. Phys. Chem. C **118**, 10963–10971 (2014)
34. T.Y. Ma, J.L. Cao, G.S. Shao, X.J. Zhang, Z.Y. Yuan, Hierarchically structured squama-like cerium-doped titania: synthesis, photoactivity, and catalytic CO oxidation. J. Phys. Chem. C **113**, 16658–16667 (2009)
35. X.J. Zhang, T.Y. Ma, Z.Y. Yuan, Titania-phosphonate hybrid porous materials: preparation, photocatalytic activity and heavy metal ion adsorption. J. Mater. Chem. **18**, 2003–2010 (2008)
36. Y.P. Zhu, T.Y. Ma, T.Z. Ren, J. Li, G.H. Du, Z.Y. Yuan, Highly dispersed photoactive zinc oxide nanoparticles on mesoporous phosphonated titania hybrid. Appl. Catal. B **156–157**, 44–55 (2014)
37. L.M. Yang, G.Y. Fang, J. Ma, E. Ganz, S.S. Han, Band gap engineering of paradigm MOF-5. Cryst. Growth Des. **4**, 2532–2541 (2014)
38. J.L. Wang, C. Wang, W. Lin, Metal–organic frameworks for light harvesting and photocatalysis. ACS Catal. **2**, 2630–2640 (2012)

39. C.G. Silva, I. Luz, F.X.L. Xamena, A. Corma, H. García, Water stable Zr-benzenedicarboxylate metal–organic frameworks as photocatalysts for hydrogen generation. Chem. Eur. J. **16**, 11133–11138 (2010)

40. C. Wang, Z. Xie, K.E. Krafft, W. Lin, Doping metal–organic frameworks for water oxidation, carbon dioxide reduction, and organic photocatalysis. J. Am. Chem. Soc. **133**, 13445–13454 (2011)

41. D. Sun, Y. Fu, W. Liu, L. Ye, D. Wang, L. Yang, X. Fu, Z. Li, Studies on photocatalytic CO_2 reduction over NH_2-Uio-66(Zr) and its derivatives: towards a better understanding of photocatalysis on metal–organic frameworks. Chem. Eur. J. **19**, 14279–14285 (2013)

42. A. Fateeva, P.A. Chater, C.P. Ireland, A.A. Tahir, Y.Z. Khimyak, P.V. Wiper, J.R. Darwent, M.J. Rosseinsky, A water-stable porphyrin-based metal–organic framework active for visible-light photocatalysis. Angew. Chem. Int. Ed. **51**, 7440–7444 (2012)

43. K. Barthelet, D. Riou, M. Nogues, G. Férey, Synthesis, structure, and magnetic properties of two new vanadocarboxylates with three-dimensional hybrid frameworks. Inorg. Chem. **42**, 1739–1743 (2003)

44. B. O'Regan, M. Grätzel, A low-cost, high-efficiency solar cell based on dye-sensitized colloidal TiO_2 films. Nature **353**, 737–740 (1991)

45. J. Burschka, N. Pellet, S.J. Moon, R. Humphry-Baker, P. Gao, M.K. Nazeeruddin, M. Grätzel, Sequential deposition as a route to high-performance perovskite-sensitized solar cells. Nature **499**, 316–320 (2013)

46. T.Y. Ma, Y.S. Wei, T.Z. Ren, L. Liu, Q. Guo, Z.Y. Yuan, Hexagonal mesoporous titanium tetrasulfonates with large conjugated hybrid framework for photoelectric conversion. ACS Appl. Mater. Interfaces **2**, 3563–3571 (2010)

47. K. Hanson, M.K. Brennaman, H. Luo, C.R.K. Glasson, J.J. Concepcion, W. Song, T.J. Meyer, Photostability of phosphonate-derivatized, Ru-II polypyridyl complexes on metal oxide surfaces. ACS Appl. Mater. Interfaces **4**, 1462–1469 (2012)

48. T.P. Brewster, S.J. Konezny, S.W. Sheehan, L.A. Martini, C.A. Schmuttenmaer, V.S. Batista, R.H. Crabtree, Hydroxamate anchors for improved photoconversion in dye-sensitized solar cells. Inorg. Chem. **52**, 6752–6764 (2013)

49. R. Luschtinetz, J. Frenzel, T. Milek, G. Seifert, Adsorption of phosphonic acid at the TiO_2 anatase (101) and rutile (110) surfaces. J. Phys. Chem. C **113**, 5730–5740 (2009)

50. K.R. Mulhern, A. Orchard, D.F. Watson, M.R. Detty, Influence of surface-attachment functionality on the aggregation, persistence, and electron-transfer reactivity of chalcogenorhodamine dyes on TiO_2. Langmuir **28**, 7071–7082 (2012)

51. K. Hanson, M.K. Brennaman, A. Ito, H. Luo, W. Song, K.A. Parker, R. Ghosh, M.R. Norris, C.R.K. Glasson, J.J. Concepcion, R. Lopez, T.J. Meyer, Structure-property relationships in phosphonate-derivatized, Ru-II polypyridyl dyes on metal oxide surfaces in an aqueous environment. J. Phys. Chem. C **116**, 14837–14847 (2012)

52. D.G. Brown, P.A. Schauer, J. Borau-Garcia, B.R. Fancy, C.P. Berlinguette, Stabilization of ruthenium sensitizers to TiO_2 surfaces through cooperative anchoring groups. J. Am. Chem. Soc. **135**, 1692–1695 (2013)

53. S. Rühle, M. Shalom, A. Zaban, Quantum-dot-sensitized solar cells. ChemPhysChem **11**, 2290–2304 (2010)

54. I. Robel, V. Subramanian, M. Kuno, P.V. Kamat, Quantum dot solar cells. Harvesting light energy with CdSe nanocrystals molecularly linked to mesoscopic TiO_2 films. J. Am. Chem. Soc. **128**, 2385–2393 (2006)

55. P. Ardalan, T.P. Brennan, H. Lee, J.R. Bakke, I.K. Ding, M.D. McGehee, S.F. Bent, Effects of self-assembled monolayers on solid-state CdS quantum dot sensitized solar cells. ACS Nano **3**, 1495–1504 (2011)

56. R.S. Dibbell, D.G. Youker, D.F. Watson, Excited-state electron transfer from CdS quantum dots to TiO_2 nanoparticles via molecular linkers with phenylene bridges. J. Phys. Chem. C **113**, 18643–18651 (2009)

57. R.S. Dibbell, D.F. Watson, Distance-dependent electron transfer in tethered assemblies of CdS quantum dots and TiO_2 nanoparticles. J. Phys. Chem. C **113**, 3139–3149 (2009)

58. Q. Li, R. He, J.O. Jensen, N.J. Bjerrum, Approaches and recent development of polymer electrolyte membranes for fuel cells operating above 100 °C. Chem. Mater. **15**, 4896–4915 (2003)
59. G.K.H. Shimizu, R. Vaidhyanathan, J.M. Taylor, Phosphonate and sulfonate metal organic frameworks. Chem. Soc. Rev. **38**, 1430–1449 (2009)
60. J.M. Taylor, R.K. Mah, I.L. Moudrakovski, C.I. Ratcliffe, R. Vaidhyanathan, G.K.H. Shimizu, Facile proton conduction via ordered water molecules in a phosphonate metal–organic framework. J. Am. Chem. Soc. **132**, 14055–14057 (2010)
61. S. Kim, K.W. Dawson, B.S. Gelfand, J.M. Taylor, G.K.H. Shimizu, Enhancing proton conduction in a metal–organic framework by isomorphous ligand replacement. J. Am. Chem. Soc. **135**, 963–966 (2013)
62. X.Q. Liang, F. Zhang, W. Feng, X.Q. Zou, C.J. Zhao, H. Na, C. Liu, F.X. Sun, G.S. Zhu, From metal–organic framework (MOF) to MOF-polymer composite membrane: enhancement of low-humidity proton conductivity. Chem. Sci. **4**, 983–992 (2013)
63. A. Bhaumik, S. Inagaki, Mesoporous titanium phosphate molecular sieves with ion-exchange capacity. J. Am. Chem. Soc. **123**, 691–696 (2001)
64. T.Y. Ma, L. Liu, Q.F. Deng, X.Z. Lin, Z.Y. Yuan, Increasing the H^+ exchange capacity of porous titanium phosphonate materials by protecting defective P–OH groups. Chem. Common. **47**, 6015–6017 (2011)
65. F. Costantino, A. Donnadio, M. Casciola, Survey on the phase transitions and their effect on the ion-exchange and on the proton-conduction properties of a flexible and robust Zr phosphonate coordination polymer. Inorg. Chem. **51**, 6992–7000 (2012)
66. T.L. Kinnibrugh, A.A. Ayi, V.I. Bakhmutov, J. Zoń, A. Clearfield, Probing structural changes in a phosphonate-based metal–organic framework exhibiting reversible dehydration. Cryst. Growth Des. **13**, 2973–2981 (2013)
67. M. Sadakiyo, T. Yamada, H. Kitagawa, Rational designs for highly proton-conductive metal–organic frameworks. J. Am. Chem. Soc. **131**, 9906–9907 (2009)
68. E. Pardo, C. Train, G. Gontard, K. Boubekeur, O. Fabelo, H. Liu, B. Dkhil, F. Lloret, K. Nakagawa, S. Ohkoshi, M. Verdaguer, High proton conduction in a chiral ferromagnetic metal-organic quartz-like framework. J. Am. Chem. Soc. **133**, 15328–15331 (2011)
69. A. Shigematsu, T. Yamada, H. Kitagawa, Wide control of proton conductivity in porous coordination polymers. J. Am. Chem. Soc. **133**, 2034–2036 (2011)
70. M.G. Goesten, J. Juan-Alcañiz, E.V. Ramos-Fernandez, K. Sai Sankar Gupta, E. Stavitski, H. Bekkum, J. Gascon, F. Kapteijn, Sulfation of metal–organic frameworks: opportunities for acid catalysis and proton conductivity. J. Catal. **281**, 177–187 (2011)
71. V.G. Ponomareva, K.A. Kovalenko, A.P. Chupakhin, D.N. Dybtsev, E.S. Shutova, V.P. Fedin, Imparting high proton conductivity to a metal–organic framework material by controlled acid impregnation. J. Am. Chem. Soc. **134**, 15640–15643 (2012)
72. J.K. Sun, Q. Xu, Functional materials derived from open framework templates/precursors: synthesis and applications. Energy Environ. Sci. **7**, 2071–2100 (2014)
73. P. Zhang, F. Sun, Z.H. Xiang, Z.G. Shen, J. Yun, D.P. Cao, ZIF-derived in situ nitrogen-doped porous carbons as efficient metal-free electrocatalysts for oxygen reduction reaction. Energy Environ. Sci. **7**, 442–450 (2014)
74. F. Afsahi, H. Vinh-Thang, S. Mikhailenko, S. Kaliaguine, Electrocatalyst synthesized from metal organic frameworks. J. Power Sources **239**, 415–423 (2013)
75. A. Dutta, A.K. Patra, H. Uyama, A. Bhaumik, Template-free synthesis of a porous organic-inorganic hybrid tin(IV) phosphonate and its high catalytic activity for esterification of free fatty acids. ACS Appl. Mater. Interfaces **5**, 9913–9917 (2013)
76. M. Pramanik, A. Bhaumik, Organic-inorganic hybrid supermicroporous iron(III) phosphonate nanoparticles as an efficient catalyst for the synthesis of biofuels. Chem. Eur. J. **19**, 8507–8514 (2013)
77. X.Z. Lin, Z.Y. Yuan, Synthesis of mesoporous zirconium organophosphonate solid-acid catalysts. Eur. J. Inorg. Chem. **2012**(16), 2661–2664 (2012)

78. M. Pramanik, M. Nandi, H. Uyama, A. Bhaumik, Organic-inorganic hybrid tinphosphonate material with mesoscopic void spaces: an excellent catalyst for the radical polymerization of styrene. Catal. Sci. Technol. **2**, 613–620 (2012)

79. A. Dutta, M. Pramanik, A.K. Patra, M. Nandi, H. Uyama, A. Bhaumik, Hybrid porous tin(IV) phosphonate: an efficient catalyst for adipic acid synthesis and a very good adsorbent for CO_2 uptake. Chem. Commun. **48**, 6738–6740 (2012)

80. A.A.G. Shaikh, S. Sivaram, Organic carbonates. Chem. Rev. **96**, 951–976 (1996)

81. J.L. Song, Z.F. Zhang, S.Q. Hu, T.B. Wu, T. Jiang, B.X. Han, MOF-5/n-Bu4NBr: an efficient catalyst system for the synthesis of cyclic carbonates from epoxides and CO_2 under mild conditions. Green Chem. **11**, 1031–1036 (2009)

82. D.A. Yang, H.Y. Cho, J. Kim, S.T. Yang, W.S. Ahn, CO_2 capture and conversion using Mg-MOF-74 prepared by a sonochemical method. Energy Environ. Sci. **5**, 6465–6473 (2012)

83. Y.W. Ren, Y.C. Shi, J.X. Chen, S.R. Yang, C.R. Qi, H.F. Jiang, Ni(salphen)-based metal–organic framework for the synthesis of cyclic carbonates by cycloaddition of CO_2 to epoxides. RSC Adv. **3**, 2167–2170 (2013)

84. D.W. Feng, W.C. Chung, Z.W. Wei, Z.Y. Gu, H.L. Jiang, Y.P. Chen, D.J. Darensbourg, H.C. Zhou, Construction of ultrastable porphyrin Zr metal–organic frameworks through linker elimination. J. Am. Chem. Soc. **135**, 17105–17110 (2013)

85. J. Kim, S.N. Kim, H.G. Jang, G. Seo, W.S. Ahn, CO_2 cycloaddition of styrene oxide over MOF catalysts. Appl. Catal. A **453**, 175–180 (2013)

86. Y.P. Zhu, Y.L. Liu, T.Z. Ren, Z.Y. Yuan, Mesoporous nickel phosphate/phosphonate hybrid microspheres with excellent performance for adsorption and catalysis. RSC Adv. **4**, 16018–16021 (2014)

87. Y.P. Zhu, T.Z. Ren, Z.Y. Yuan, Hollow cobalt phosphonate spherical hybrid as high-efficiency Fenton catalyst. Nanoscale **6**, 11395–11402 (2014)

88. Y.K. Hwang, D.Y. Hong, J.S. Chang, S.H. Jhung, Y.K. Seo, J. Kim, A. Vimont, M. Daturi, C. Serre, G. Férey, Amine grafting on coordinatively unsaturated metal centers of MOFs: consequences for catalysis and metal encapsulation. Angew. Chem. Int. Ed. **47**, 4144–4148 (2008)

89. J. Gascon, U. Aktay, M.D. Hernandez-Alonso, G.P.M. van Klink, F. Kapteijn, Amino-based metal–organic frameworks as stable, highly active basic catalysts. J. Catal. **261**, 75–87 (2009)

90. N.V. Maksimchuk, K.A. Kovalenko, V.P. Fedinm, O.A. Kholdeeva, Heterogeneous selective oxidation of alkenes to alpha, beta-unsaturated ketones over coordination polymer MIL-101. Adv. Synth. Catal. **352**, 2943–2948 (2010)

91. T.Y. Ma, Z.Y. Yuan, Functionalized periodic mesoporous titanium phosphonate monoliths with large ion exchange capacity. Chem. Commun. **46**, 2325–2327 (2010)

92. T.Y. Ma, Z.Y. Yuan, Periodic mesoporous titanium phosphonate spheres for high dispersion of CuO nanoparticles. Dalton Trans. **39**, 9570–9578 (2010)

93. J.L. Cao, Y. Wang, X.L. Yu, S.R. Wang, S.H. Wu, Z.Y. Yuan, Mesoporous CuO-Fe2O3 composite catalysts for low-temperature carbon monoxide oxidation. Appl. Catal. B **26**, 26–34 (2008)

94. J.L. Cao, G.S. Shao, Y. Wang, Y.P. Liu, Z.Y. Yuan, CuO catalysts supported on attapulgite clay for low-temperature CO oxidation. Catal. Commun. **9**, 2555–2559 (2008)

95. H.P. Perry, J. Law, J. Zon, A. Clearfield, Porous zirconium and tin phosphonates incorporating 2,2′-bipyridine as supports for palladium nanoparticles. Micropor. Mesopor. Mater. **149**, 172–180 (2012)

96. X.Y. Liu, A.Q. Wang, X.F. Yang, T. Zhang, C.Y. Mou, D.S. Su, J. Li, Synthesis of thermally stable and highly active bimetallic Au-Ag nanoparticles on inert supports. Chem. Mater. **21**, 410–418 (2009)

97. J. Canivet, S. Aguado, Y. Schuurman, D. Farrusseng, MOF-supported selective ethylene dimerization single-site catalysts through one-pot postsynthetic modification. J. Am. Chem. Soc. **135**, 4195–4198 (2013)

98. F. Schroeder, D. Esken, M. Cokoja, M.W.E. van den Berg, O.I. Lebedev, G. Van Tendeloo, B. Walaszek, G. Buntkowsky, H.H. Limbach, B. Chaudret, R.A. Fischer, Ruthenium nanoparticles inside porous [Zn₄O(bdc)₃] by hydrogenolysis of adsorbed [Ru(cod)(cot)]: a solid-state reference system for surfactant-stabilized ruthenium colloids. J. Am. Chem. Soc. **130**, 6119–6130 (2008)

99. M. Meilikhov, K. Yusenko, D. Esken, S. Turner, G. Van Tendeloo, R.A. Fischer, Metals@MOFs loading MOFs with metal nanoparticles for hybrid functions. Eur. J. Inorg. Chem. **2010**, 3701–3714 (2010)

100. A. Henschel, K. Gedrich, R. Kraehnert, S. Kaskel, Catalytic properties of MIL-101. Chem. Commun. **2008**, 4192–4194 (2008)

101. Y. Pan, B. Yuan, Y. Li, D. He, Multifunctional catalysis by Pd@MIL-101: one-step synthesis of methyl isobutyl ketone over palladium nanoparticles deposited on a metal–organic framework. Chem. Commun. **46**, 2280–2282 (2010)

102. H. Liu, Y. Liu, Y. Li, Z. Tang, H. Jiang, Metal–organic framework supported gold nanoparticles as a highly active heterogeneous catalyst for aerobic oxidation of alcohols. J. Phys. Chem. C **114**, 13362–13369 (2010)

103. Y.K. Park, S.B. Choi, H.J. Nam, D.Y. Jung, H.C. Ahn, K. Choi, H. Furukawad, J. Kim, Catalytic nickel nanoparticles embedded in a mesoporous metal–organic framework. Chem. Commun. **46**, 3086–3088 (2010)

104. S. Hudson, J. Cooney, E. Magner, Proteins in mesoporous silicates. Angew. Chem. Int. Ed. **47**, 8582–8594 (2008)

105. V. Lykourinou, Y. Chen, X.S. Wang, L. Meng, T. Hoang, L.J. Ming, R.L. Musselman, S.Q. Ma, Immobilization of MP-11 into a mesoporous metal–organic framework, MP-11@mesoMOF: a new platform for enzymatic catalysis. J. Am. Chem. Soc. **133**, 10382–10385 (2011)

106. X. Shi, J. Liu, C.M. Li, Q.H. Yang, Pore-size tunable mesoporous zirconium organophosphonates with chiral (L)-proline for enzyme adsorption. Inorg. Chem. **46**, 7944–7952 (2007)

107. T.Y. Ma, X.J. Zhang, Z.Y. Yuan, Hierarchical meso-/macroporous aluminum phosphonate hybrid materials as multifunctional adsorbents. J. Phys. Chem. C **113**, 12854–12862 (2009)

108. Y.P. Zhu, Y.L. Liu, T.Z. Ren, Z.Y. Yuan, Hollow manganese phosphonate microspheres with hierarchical porosity for efficient adsorption and separation. Nanoscale **6**, 6627–6636 (2014)

109. P. Horcajada, C. Serre, M. Vallet-Regí, M. Sebban, F. Taulelle, G. Férey, Metal–organic frameworks as efficient materials for drug delivery. Angew. Chem. Int. Ed. **45**, 5974–5978 (2006)

110. P. Horcajada, T. Chalati, C. Serre, B. Gillet, C. Sebrie, T. Baati, J.F. Eubank, D. Heurtaux, P. Clayette, C. Kreuz, J.S. Chang, Y.K. Hwang, V. Marsaud, P.N. Bories, L. Cynober, S. Gil, G. Férey, P. Couvreur, R. Gref, Porous metal-organic-framework nanoscale carriers as a potential platform for drug delivery and imaging. Nat. Mater. **9**, 172–178 (2010)

111. X. Shi, J.P. Li, Y. Tang, Q.H. Yang, pH-Sensitive mesoporous zirconium diphosphonates for controllable colon-targeted delivery. J. Mater. Chem. **20**, 6495–6504 (2010)

112. Y. Tang, Y.B. Ren, X. Shi, Bifunctional mesoporous zirconium phosphonates for delivery of nucleic acids. Inorg. Chem. **52**, 1388–1397 (2013)

113. K.M.L. Taylor-Pashow, J.D. Rocca, Z. Xie, S. Tran, W. Lin, Postsynthetic modifications of iron-carboxylate nanoscale metal–organic frameworks for imaging and drug delivery. J. Am. Chem. Soc. **131**, 14261–14263 (2009)

114. W.T. Al-Jamal, K. Kostarelos, Liposomes: from a clinically established drug delivery system to a nanoparticle platform for theranostic nanomedicine. Acc. Chem. Res. **44**, 1094–1104 (2011)

115. Y. Namiki, T. Fuchigami, N. Tada, R. Kawamura, S. Matsunuma, Y. Kitamoto, M. Nakagawa, Nanomedicine for cancer: lipid-based nanostructures for drug delivery and Monitoring. Acc. Chem. Res. **44**, 1080–1093 (2011)

116. C. Wang, D.M. Liu, W.B. Lin, Metal–organic frameworks as a tunable platform for designing functional molecular materials. J. Am. Chem. Soc. **135**, 13222–13234 (2013)

117. A.K. Gupta, M. Gupta, Synthesis and surface engineering of iron oxide nanoparticles for biomedical applications. Biomaterials **26**, 3995–4021 (2005)

118. E. Duguet, S. Vasseur, S. Mornet, J.M. Devoiselle, Magnetic nanoparticles and their applications in medicine. Nanomedicine **1**, 157–168 (2006)
119. L. Lartigue, C. Innocenti, T. Kalaivani, A. Awwad, D.M. Sanchez, Y. Guari, J. Larionova, C. Guerin, J.L.G. Montero, V. Barragan-Montero, P. Arosio, A. Lascialfari, D. Gatteschi, C. Sangregorio, Water-dispersible sugar-coated iron oxide nanoparticles. an evaluation of their relaxometric and magnetic hyperthermia properties. J. Am. Chem. Soc. **133**, 10459–10472 (2011)
120. H. Dong, J. Huang, R.R. Koepsel, P. Ye, A.J. Russell, K. Matyjaszewski, Recyclable antibacterial magnetic nanoparticles grafted with quaternized poly(2-(dimethylamino)ethyl methacrylate) brushes. Biomacromolecules **12**, 1305–1311 (2011)
121. C. Tudisco, V. Oliveri, M. Cantarella, G. Vecchio, G.G. Condorelli, Cyclodextrin anchoring on magnetic Fe_3O_4 nanoparticles modified with phosphonic linkers. Eur. J. Inorg. Chem. **32**, 5323–5331 (2012)
122. H.K. Chang, F.N. Ishikawa, R. Zhang, R. Datar, R.J. Cote, M.E. Thompson, C.W. Zhou, Rapid, label-free, electrical whole blood bioassay based on nanobiosensor systems. ACS Nano **5**, 9883–9891 (2011)
123. K. Wang, Z. Tang, C.J. Yang, Y. Kim, X. Fang, W. Li, Y. Wu, C.D. Medley, Z. Cao, J. Li, P. Colon, H. Lin, W. Tan, Molecular engineering of DNA: molecular beacons. Angew. Chem. Int. Ed. **48**, 856–870 (2009)
124. K.E. Uhrich, S.M. Cannizzaro, R.S. Langer, K.M. Shakesheff, Polymeric systems for controlled drug release. Chem. Rev. **99**, 3181–3198 (1999)
125. H. Li, T.Y. Ma, D.M. Kong, Z.Y. Yuan, Mesoporous phosphonate-TiO_2 nanoparticles for simultaneous bioresponsive sensing and controlled drug release. Analyst **4**, 1084–1090 (2013)
126. E.Z. Lee, Y.S. Jun, W.H. Hong, A. Thomas, M.M. Jin, Cubic mesoporous graphitic carbon(IV) nitride: an all-in-one chemosensor for selective optical sensing of metal ions. Angew. Chem. Int. Ed. **49**, 9706–9710 (2010)
127. Y.P. Zhu, T.Y. Ma, T.Z. Ren, Z.Y. Yuan, Mesoporous cerium phosphonate nanostructured hybrid spheres as label-free Hg^{2+} fluorescent probes. ACS Appl. Mater. Interfaces **6**, 16344–16351 (2014)
128. B. Chen, L. Wang, F. Zapata, G. Qian, E.B. Lobkovsky, A luminescent microporous metal–organic framework for the recognition and sensing of anions. J. Am. Chem. Soc. **130**, 6718–6719 (2008)
129. X. Zhu, H. Zheng, X. Wei, Z. Lin, L. Guo, B. Qiu, G. Chen, Metal–organic framework (MOF): a novel sensing platform for biomolecules. Chem. Commun. **49**, 1276–1278 (2013)
130. F. Luo, S.R. Batten, Metal–organic framework (MOF): lanthanide(III)-doped approach for luminescence modulation and luminescent sensing. Dalton Trans. **39**, 4485–4488 (2010)

Chapter 6
Summary and Outlook

Abstract A summary of the synthesis, properties and applications of mesoporous non-siliceous organic–inorganic hybrid materials is provided herein. With the use of organophosphonic/sulfonic/carboxylic acids and their derivatives (salts, esters) as coupling molecules, mesoporous metal phosphonate/sulfonate/carboxylate materials with a homogeneous distribution of considerable organic functional groups in the hybrid frameworks could be achieved. A variety of methods have been developed to successfully incorporate mesoporosity into the hybrid networks and control the resultant micro-/macro-morphologies. Mesoporous non-siliceous organic–inorganic materials have found diverse potential applications across broad ranges in adsorption and separation, catalysis, sustainable energy conversion and storage, and biomaterials. Nonetheless, there are still unclear areas that deserve research fellows to imagine and explore.

Keywords Mesoporous materials · Organic–inorganic hybrid · Non-siliceous hybrid · Perspectives

Mesoporous metal carboxylates, phosphonates, and sulfonates, as the three most fascinating members of non-siliceous hybrids with alternative organic–inorganic frameworks, have attracted great research interest in the past decades due to their outstanding physicochemical properties. In order to improve the accessibility of the pores, a template-free methodology via enlarging the organic linkages has been proved to be facile and effective, but restricted. The soft-templating approach has been realized to synthesize mesoporous hybrids with a uniform pore width, high surface area, and even hierarchical porosity. The control over the mesophase symmetry, the pore sizes, and the crystallinity of the pore walls is relevant to the structure of the surfactant molecules and the synthesis method. Morphological adjustment through a variety of fabrication techniques is feasible, which endows them with capabilities in diverse production fields. The synthesized mesoporous non-silica-based hybrid materials could be utilized as efficient host solids for the adsorption and separation of gas, liquid, heavy metal ions, as well as organic

© The Author(s) 2015
Y.-P. Zhu and Z.-Y. Yuan, *Mesoporous Organic-Inorganic Non-Siliceous Hybrid Materials*, SpringerBriefs in Molecular Science,
DOI 10.1007/978-3-662-45634-7_6

constituents. They were also useful for eco-friendly photocatalysis and solar cells under simulated solar light irradiation, and biomaterials for enzymatic engineering, drug delivery, and medical diagnosis. Further functionalization of the mesoporous hybrids could make them oxidation and acid catalysts, both with impressive performances in the areas of sustainable energy and environment.

As compared with the counterparts of phosphonates and carboxylates, reports with regard to mesoporous metal sulfonates are relatively rare. This is mainly due to the weaker combination between metal linkers and organosulfonic groups, which leads to a less robust framework. However, besides applications in adsorption/separation and photoelectrochemistry, the weaker ligating nature of sulfonic bridging units predisposes the network to certain degrees of flexibility, namely a dynamics material. Correspondingly, the resultant materials can be used to selectively adsorb and detect metal ions and small molecules. The reversible insertion/desertion of Li^+ and protons through the pores and elastic network can also be envisioned. Although this aspect is not extensively studied, the intrinsic porosity within the electrically conductive hybrid materials remains largely unknown but is worthy of research efforts.

The strong binding ability of phosphonic acids usually leads to dense layered architectures of metal phosphonates. Correspondingly, phosphonate-based MOFs have distinct differences from the carboxylate- and sulfonate-based counterparts, such as the relatively high thermal and chemical stability and extremely low solubility. On the contrary, these typically render metal phosphonates difficult to obtain crystalline phases with determined structures. Developments of high-throughput hydrothermal or solvothermal techniques and advances in powder XRD modeling and refinement will significantly increase the number of structurally characterized phosphonate-based MOFs, and it will be exciting to watch this field as it develops.

The exploration of synthetic methodologies and extended applications of mesoporous non-siliceous hybrid materials remains promising and valuable. Due to the complexity of the interactions between organic groups and metallic centers, the achievement of intentional control over the pore size, pore channel regularity and mesophase is still a challenge. The practical value of mesoporous non-siliceous hybrids is limited by their relatively poor thermal and hydrothermal stabilities as compared with those of silica-based materials. The effort for the enhancement of the thermal and hydrothermal stability of metal phosphonate materials is of great importance, and the crystalline pore wall is first in line. However, it remains contradictory and challenging to achieve the high crystallization and well-structured hierarchical porosity (especially micro-/mesoporosity) simultaneously. The coordination rate between the inorganic units and organic moieties determines the nucleation kinetics and thus the crystal growth. "Crown-mediated controlled-release" methodology might provide a new route in the rational design of crystalline metal phosphonates with well-defined porous structures. Considering the intimate relation of phosphonates versus phosphates, carboxylates versus carbonates, and sulfonates versus sulfates, non-siliceous materials may act as precursors in preparing inorganic courterparts with interesting structures and novel properties, thus improving the performances of phosphates in catalysis, photoelectricity, and secondary batteries.

Overall, the development in organic–inorganic non-siliceous hybrid materials signifies a pivotal step toward exploring and finding new multifunctional hybrid materials and has significantly expanded the application ranges. Nowadays, the molecular approaches of solid-state chemistry and organic synthesis have reached a high level of sophistication. As a consequence, original hybrids can be designed through the synthesis of new hybrid nanosynthons, allowing for the coding of hybrid assemblies presenting a spatial ordering at different length scales. Particularly, the synthesis through the simultaneous use of self-assembly processes together with external factors, such as electrical or magnetic fields, or even through the use of strong compositional flux variations of the reagents during the synthesis is a worthwhile area to explore. The increasing interest in the field of functional hybrid materials will be amplified in the future by the growing interest of materials scientists, chemists, and biologists to fully exploit this opportunity for discovering materials and devices benefiting from the best of the two realms: inorganic and organic. Besides of their high versatility that offers a wide range of possibilities in terms of physicochemical properties and shaping, hybrid nanocomposites present the paramount advantage of facilitating both integration and miniaturization of the devices, thereby offering a prospect of promising applications in a variety of fields including catalysis, photoelectrochemistry, ionics, mechanics, seperation, functional and protective coatings, sensors, biology, and medicine. Finally, the explosion of new strategies that we are presently witnessing for rational design of innovative hybrid materials allows us to dream of further challenging steps in creating intelligent materials. Therefore, it is wonderful to envision the possibility of building in the future advanced materials that will respond to external stimuli, intellectively adapt, self-replicate, self-repair or self-destroy at the end of their useful lifetime.